# FOCUS ON MIDDLE SCHOOL

Grades 5-8

## Teacher's Manual
### 3rd Edition

Rebecca W. Keller, PhD

# Real Science-4-Kids

Cover design:   David Keller, PhD
Opening page:   David Keller, PhD

Copyright © 2019 Gravitas Publications Inc.

All rights reserved. No part of this publication may be reproduced, stored in a retrieval system, or transmitted, in any form or by any means, electronic, mechanical, photocopying, recording, or otherwise, without prior written permission from the publisher. No part of this book may be reproduced in any manner whatsoever without written permission.

Focus On Middle School Biology Teacher's Manual—3rd Edition
ISBN  978-1-941181-50-8

Published by Gravitas Publications Inc.
www.gravitaspublications.com
www.realscience4kids.com

## A Note from the Author

This curriculum is designed to engage middle school level students in further exploration of the scientific discipline of biology. The *Focus On Middle School Biology Student Textbook—3rd Edition* and the accompanying *Laboratory Notebook* together provide students with basic science concepts needed for developing a solid framework for real science investigation into biology.

The *Laboratory Notebook* contains 44 experiments—two experiments for each chapter of the *Student Textbook*. These experiments allow students to expand on concepts presented in the *Student Textbook* and develop the skills needed for using the scientific method. This *Teacher's Manual* will help you guide students through the laboratory experiments.

There are several sections in each chapter of the *Laboratory Notebook*. The section called *Think About It* provides questions to help students develop critical thinking skills and spark their imagination. The *Experiment* section provides students with a framework to explore concepts presented in the *Student Textbook*. In the *Conclusions* section students draw conclusions from the observations they have made during the experiment. A section called *Why?* provides a short explanation of what students may or may not have observed. And finally, in each chapter an additional experiment is presented in *Just For Fun*.

The experiments take up to 1 hour. Materials needed for each experiment are listed on the following pages and also at the beginning of each experiment.

Enjoy!

*Rebecca W. Keller, PhD*

# Materials at a Glance

| Experiment 1 | Experiment 2 | Experiment 3 | Experiment 4 | Experiment 5 |
|---|---|---|---|---|
| pencil and eraser<br>Objects chosen by students, such as:<br>  rubber ball<br>  cotton ball<br>  orange<br>  banana<br>  apple<br>  paper<br>  sticks<br>  leaves<br>  rocks<br>  grass<br>  Legos<br>  building blocks<br>  other objects<br>**Optional**<br>several sheets of paper | plastic petri dishes[1]<br>dehydrated agar powder[2]<br>distilled water<br>K-12 safe E. coli bacterial culture[3]<br>inoculation loop[4]<br>candle or gas flame<br>cooking pot<br>mixing spoon<br>oven mitt or pot holder<br>measuring spoons<br>measuring cup<br>black permanent marker<br>red marker<br>rubber gloves, 2 pairs | microscope with 4X, 10X, and 40X objective lenses; 100X objective lens recommended but not required. (See beginning of chapter for purchasing info.)<br>glass microscope slides[5]<br>glass microscope cover slips[6]<br>immersion oil (if using 100X objective lens)[7]<br>Samples:<br>  piece of paper with lettering<br>  strands of hair<br>  droplet of blood<br>  insect wing | tincture of iodine [VERY POISONOUS—DO NOT ALLOW STUDENTS TO EAT any food items that have iodine on them]<br>bread, 1 slice<br>timer<br>wax paper<br>marking pen<br>cup<br>refrigerator<br>a green vegetable<br>one or more other vegetables or fruits | pencil<br>colored pencils/crayons<br>student-selected materials for model cell |

| | | | | Experiment 6 |
|---|---|---|---|---|
| | | | | dehydrated agar[2]<br>distilled water<br>cooking pot<br>measuring spoons<br>measuring cup<br>cup<br>plastic petri dishes (20)[1]<br>cotton swabs<br>permanent marker<br>oven mitt or pot holder |

| Experiment 7 | Experiment 8 | Experiment 9 | Experiment 10 | Experiment 11 |
|---|---|---|---|---|
| microscope with a 10X objective<br>microscope depression slides[8]<br>10 or more eyedroppers<br>fresh pond water or water mixed with soil<br>protozoa study kit[9] (must be used within 1-2 days of arrival)<br>methyl cellulose[10]<br>measuring cup and measuring spoons<br>baker's yeast<br>distilled water<br>Eosin Y stain[11] | agar powder[2]<br>distilled water<br>cooking pot<br>measuring spoons<br>measuring cup<br>plastic petri dishes[1]<br>permanent marker<br>oven mitt or pot holder<br>jar with lid (big enough to hold 235 ml (about 1 cup) liquid<br>1 slice of bread, preferably preservative free<br>small clear plastic bag<br>white vinegar<br>bleach<br>borax<br>mold or mildew cleaner<br>1-2 pairs rubber gloves | colored pencils<br>handheld magnifying glass<br>field notebook (blank or faintly lined pages)<br>backpack, water, snacks<br>2 plant pots<br>potting soil and water<br>corn seeds, 8 or more with packet<br>bean seeds, 8 or more with packet<br>**Optional**<br>field guide to the plants book<br>iPad, camera, or smartphone with camera<br>plant identification app: do some online research to find the best app to use with a specific mobile device | plant with at least 6 flat, green leaves (a tree may be used)<br>lightweight cardboard or construction paper—enough to cut out 6 pieces that are bigger than a leaf<br>scissors<br>tape<br>2 small jars<br>marking pen<br>4 or more plant pots<br>potting soil<br>bean seeds (12 or more) | microscope with 4X, 10X, and 40X objective lenses; a 100X objective lens is recommended<br>glass microscope slides (plain)[5]<br>glass coverslips[6]<br>immersion oil (if using 100X objective lens)[7]<br>water<br>eyedropper<br>sharp knife<br>toothpick<br>colored pencils<br>Samples:<br>  raw celery stalk with leaves<br>  raw carrot<br>  a large leaf<br>  other plant parts: students' choice<br>3 or more small jars<br>several fresh white carnation flowers<br>food coloring |

As of this writing the following materials are available from http://www.hometrainingtools.com/

1. A stack of 20 petri dishes: petri-dishes-plastic-20-pk/p/BE-PETRI20/
2. Nutrient-agar-8-g-dehydrated/p/CH-AGARN08/
3. Escherichia-coli-bacteria/p/LD-ESCHCOL/
4. Inoculating-needle-looped-end/p/BE-INOCUL/
5. Glass microscope slides: MS-SLIDP72 or MS-SLIDEPL
6. Glass microscope cover slip: MS-SLIDCV
7. Immersion oil: MI-IMMOIL
8. Glass Depression Slides, MS-SLIDC72 or MS-SLIDC12
9. Basic Protozoa Set, LD-PROBASC
10. Methyl Cellulose, CH-METHCEL
11. Eosin Y, CH-EOSIN

(Or search by the name of the item needed)

| Experiment 12 | Experiment 13 | Experiment 14 | Experiment 15 | Experiment 16 |
|---|---|---|---|---|
| several fresh vegetable scraps such as: carrot top, lettuce leaves or the root end of a head of lettuce, red beet top, turnip top, garlic bulb, onion bulb, scallions, either or both ends of a zucchini squash or cucumber, basil leaves with stem, potato (piece or peeling with eyes), or other vegetables of students' choice<br>knife<br>toothpicks<br>several small glass jars or small drinking glasses<br>colored pencils or pens<br>several plant pots<br>potting soil<br>water<br>**Optional**<br>existing or new field notebook<br>garden trowel or spoon | toothpicks or cotton swabs<br>glass microscope slides[1]<br>plastic pipette or eyedropper[1]<br>methylene blue solution (0.5% to 1%)[1] (iodine can be used instead—follow the same safety precautions)<br>plastic cover slip[1]<br>paper towels or tissues<br>thin rubber, vinyl, or latex gloves that are a tight fit<br>goggles or other eye protection[1]<br>microscope<br>misc. household materials to make microscope dyes<br>**Optional**<br>immersion oil[1] | **14A**<br>preserved specimens: clam, crayfish, sea star, and earthworm, (non-injected or injected)[2]<br>dissection guide for each organism[2]<br>safety goggles<br>lab apron<br>gloves<br>dissection tray<br>dissection pins<br>dissecting probe<br>forceps<br>scissors<br>scalpel<br>hand lens or magnifying glass<br>paper towels<br>water<br><br>**14B**<br>food items:<br>  sugar cube<br>  small piece of animal protein (chunk of turkey, ham, roast beef, etc.)<br>  cheese<br>  apple<br>  bread<br>  oil or butter<br>choice chamber, homemade:<br>  shallow pan, shallow cardboard box, short jar, or plastic Petri dish<br>  cardboard or paper cut into strips<br>choice chamber, purchased: available from Home Science Tools; search on "choice chamber."[1] | **15A**<br>preserved specimens: frog, shark, and perch (Specimens don't need to be injected.)<br>dissection guide for each organism<br>safety goggles<br>lab apron<br>gloves<br>dissection tray<br>dissection pins<br>dissecting probe<br>forceps<br>scissors<br>scalpel<br>hand lens or magnifying glass<br>paper towels<br>water<br><br>**15B**<br>ebird.org app (free)<br>Merlin Bird ID app (free) or other bird ID app and/or a print book field guide to the birds, such as *The Young Birder's Guide to North America*<br>smartphone or iPad with internet access and camera; or desktop or laptop computer and digital camera, if available<br>an email address<br>field notebook (existing or new)<br>pen, pencil, colored pencils<br>**Optional**<br>binoculars | **16A**<br>preserved fetal pig (doesn't need to be injected)<br>dissection guide<br>safety goggles<br>lab apron<br>gloves<br>dissection tray<br>dissection pins<br>dissecting probe<br>forceps<br>scissors<br>scalpel<br>hand lens or magnifying glass<br>paper towels<br>water<br><br>**16B**<br>smartphone, iPad, or computer with internet access and camera; or desktop or laptop computer and digital camera, if available<br>an email address<br>field notebook (an existing one or start a new one for citizen science projects)<br>**Or**<br>Local library, zoo, or natural history museum<br>field notebook (an existing one or start a new one for citizen science projects) |

1. Available from Home Science Tools: https://www.homesciencetools.com/
   Type the name of the item needed in the website search bar.

**Experiments 14-16**

Most of the supplies are available from Home Science Tools. Type the name of the item needed in the website search bar.

For preserved organisms and dissection guides search on the Home Science Tools website for "dissection specimen" and "dissection guide." Choose the organisms listed for each experiment. (At the time of this writing, Home Science Tools offers an "Animal Specimen Set of 9 with Pig" that has most of the specimens needed for Experiments 14-16) Dissection tools are also available from Home Science Tools. Search for individual tools or a dissection kit. Look for other supplies too.

https://www.homesciencetools.com/

# Materials

## Quantities Needed for All Experiments

| Equipment | Materials | Materials |
|---|---|---|
| backpack<br>choice chamber, homemade:<br>   shallow pan, shallow cardboard box,<br>   short jar, or plastic Petri dish, and<br>   cardboard or paper cut into strips<br>choice chamber, purchased: available from<br>   Home Science Tools; search on "choice<br>   chamber" *<br>cooking pot<br>cup<br>cup, measuring<br>dissecting probe *<br>dissection pins *<br>dissection tray *<br>forceps *<br>goggles, safety, or other eye protection *<br>hand lens or magnifying glass *<br>inoculation loop [4]<br>jar with lid (big enough to hold 235 ml<br>   liquid (about 1 cup)<br>jars, 5 or more small<br>jars, small glass or small drinking glasses<br>   (several)<br>knife, sharp<br>lab apron *<br>microscope with 4X, 10X, and 40X<br>   objective lenses; a 100X objective lens<br>   is recommended (see Chapter 3 for<br>   selection info & advice)<br>oven mitt or pot holder<br>plant pots (6 or more)<br>refrigerator<br>scalpel *<br>scissors<br>smartphone or iPad with internet access<br>   and camera; or desktop or laptop<br>   computer and digital camera, if<br>   available<br>spoon, mixing<br>spoons, measuring<br>timer<br>**Optional**<br>binoculars<br>field guide to plants print book/field guide<br>   to birds<br>iPad, camera, or smartphone with camera<br>library, zoo, or natural history museum in<br>   your area<br>plant identification app (do some online<br>   research to find the best app to use with<br>   a specific mobile device)<br>trowel, garden, or spoon | agar, dehydrated powder [2]<br>bleach<br>borax<br>candle (or gas stove flame)<br>cardboard, lightweight, or construction<br>   paper<br>carnation flowers, several fresh white<br>cleaner, mold or mildew<br>cotton swabs<br>E. coli bacterial culture, K-12 safe [3]<br>Eosin Y stain [11]<br>eraser<br>eyedroppers (11 or more) *<br>food coloring<br>gloves, rubber, 3-4 pairs<br>gloves, thin rubber, vinyl, or latex, that are<br>   a tight fit (several pairs)<br>immersion oil (if using 100X objective<br>   lens) [7]<br>iodine, tincture of [VERY POISONOUS—<br>   DO NOT ALLOW STUDENTS TO<br>   INGEST] *<br>leaf, large<br>marker, black permanent<br>marker, red permanent<br>methyl cellulose [10]<br>methylene blue solution (0.5% to 1%) [1]<br>   (iodine can be used instead—follow the<br>   same safety precautions) *<br>microscope cover slips, glass [6]<br>microscope cover slips, plastic *<br>microscope slides , depression [8]<br>microscope slides, plain, glass [5]<br>notebook, for field notebook, existing or<br>   new (1 or more), unlined or faint lines<br>   works best<br>paper<br>paper towels or tissues<br>pencil<br>pencils, colored, or crayons<br>petri dishes, plastic (50-60) [1]<br>pipette, plastic, or eyedropper * | plant with at least 6 flat, green leaves (a<br>   tree may be used)<br>plastic bag, small clear<br>potting soil<br>protozoa study kit [9] (must be used within<br>   1-2 days of arrival)<br>seeds, bean 20 or more with packet<br>seeds, corn, 8 or more with packet<br>tape<br>toothpicks<br>vinegar, white<br>water, distilled<br>water, fresh pond or water mixed with soil<br>wax paper |

| | | Materials, Misc. |
|---|---|---|
| | | materials, household (misc.) to make<br>   microscope dyes (students' choice)<br>materials, student-selected, to make a<br>   model cell<br>objects chosen by students, such as:<br>   rubber ball<br>   cotton ball<br>   orange<br>   banana<br>   apple<br>   paper<br>   sticks<br>   leaves<br>   rocks<br>   grass<br>   Legos<br>   building blocks<br>   other objects<br>plant parts, misc., students' choice<br>samples for microscopy:<br>   blood, droplet<br>   hair, a few strands<br>   insect wing<br>   paper, piece with lettering |

| Other | Preserved Specimens* | Foods |
|---|---|---|
| ebird.org app (free)<br>email address<br>Merlin Bird ID app (free) or other bird ID app and/or a print book field guide to the birds, such as *The Young Birder's Guide to North America* | [can use either non-injected or injected specimens]<br><br>clam<br>crayfish<br>earthworm<br>fetal pig<br>frog<br>perch<br>sea star<br>shark<br><br>dissection guide for each organism* | animal protein (chunk of turkey, ham, roast beef, etc.), small piece<br>apple<br>bread, any, 1-2 slices<br>bread, 1 slice, preferably preservative free<br>carrot, raw<br>celery stalk with leaves, raw<br>cheese<br>oil or butter<br>snacks<br>sugar cube<br>vegetable, green (student's choice)<br>vegetables or fruits (misc.), one or more<br>vegetable scraps, several fresh, such as: carrot top, lettuce leaves or the root end of a head of lettuce, red beet top, turnip top, garlic bulb, onion bulb, scallions, either or both ends of a zucchini squash or cucumber, basil leaves with stem, potato (piece or peeling with eyes), or other vegetables of students' choice<br>yeast, baker's |

As of this writing the following materials are available from http://www.hometrainingtools.com/

1. A stack of 20 petri dishes: petri-dishes-plastic-20-pk/p/BE-PETRI20/
2. Nutrient-agar-8-g-dehydrated/p/CH-AGARN08/
3. Escherichia-coli-bacteria/p/LD-ESCHCOL/
4. Inoculating-needle-looped-end/p/BE-INOCUL/
5. Glass microscope slides: MS-SLIDP72 or MS-SLIDEPL
6. Glass microscope cover slip: MS-SLIDCV
7. Immersion oil: MI-IMMOIL
8. Glass Depression Slides, MS-SLIDC72 or MS-SLIDC12
9. Basic Protozoa Set, LD-PROBASC
10. Methyl Cellulose, CH-METHCEL
11. Eosin Y, CH-EOSIN

(Or search by the name of the item needed)

* Available from Home Science Tools: https://www.homesciencetools.com/
Type the name of the item needed in the website search bar.

For preserved organisms and dissection guides search on the Home Science Tools website for "dissection specimen" and "dissection guide." Choose the organisms listed for each experiment. (At the time of this writing, Home Science Tools offers an "Animal Specimen Set of 9 with Pig" that has most of the specimens needed for Experiments 14-16) Dissection tools are also available from Home Science Tools. Search for individual tools or a dissection kit. Look for other supplies too.
https://www.homesciencetools.com/

# Contents

| | | |
|---|---|---|
| Experiment 1 | **Putting Things in Order** | 1 |
| Experiment 2 | **Using Agar Plates** | 5 |
| Experiment 3 | **Using A Light Microscope** | 9 |
| Experiment 4 | **What's in Spit?** | 15 |
| Experiment 5 | **Inside the Cell** | 18 |
| Experiment 6 | **Wash Your Hands!** | 23 |
| Experiment 7 | **Observing Protists** | 27 |
| Experiment 8 | **Moldy Growth** | 31 |
| Experiment 9 | **Identifying Plants** | 35 |
| Experiment 10 | **Take Away the Light** | 39 |
| Experiment 11 | **Seeing Inside Plants** | 44 |
| Experiment 12 | **Growing Vegetables from Scraps** | 50 |
| Experiment 13 | **Human Cheek Cells** | 54 |
| Experiment 14 | **Non-chordates** | 57 |
| Experiment 15 | **Chordates** | 64 |
| Experiment 16 | **Mammals** | 73 |

## Experiment 1
# Putting Things in Order

**Materials Needed**

- pencil and eraser

Objects chosen by students, such as:

- rubber ball
- cotton ball
- orange
- banana
- apple
- paper
- sticks
- leaves
- rocks
- grass
- Legos
- building blocks
- other objects

**Optional**

- several sheets of paper

## Objectives

In this experiment students explore categorizing objects by their features.

The objectives of this lesson are for students to:

- Explore how objects can be categorized in different ways and how to chart their data.
- Observe the difficulties of categorizing objects.

## Experiment

### I. Think About It

Read this section of the *Laboratory Notebook* with your students.

Ask questions such as the following to guide open inquiry.

- *What are some groups of objects you can think of?*
- *How would you decide which objects should go in each group?*
- *Do you think it can be helpful to you to put objects into groups? Why or why not?*
- *Do you think some objects can go into more than one group? Why or why not?*
- *Do you think it is easy or difficult to put objects in groups? Why?*
- *How do you use groups in your day-to-day life?*

### II. Experiment 1: Putting Things in Order

In this experiment, students will try to organize different objects according to their characteristics, such as shape, color, or texture. There are no "right" answers for this experiment, and the categories the students choose will vary.

Have the students read the entire experiment.

Help them collect a wide variety of objects of their choice that they will categorize.

**Objective:** Have the students write an objective. Some examples:

- *To put objects into different categories.*
- *To use categories and subcategories.*

**Hypothesis:** Have the students write a hypothesis. Some examples:

- *It will be easy to put objects in categories.*
- *Some objects will go into more than one category.*

## EXPERIMENT

❶ Have the students place the collected objects on a table and then make careful observations. Guide them to notice some features of the objects, such as color, shape, and texture. Also, discuss any common uses, for example, those used as toys or those used as writing instruments.

❷ Have the students fill in the chart provided, listing each object and a few of its characteristics. Help them to be as descriptive as possible. For example, oranges can be described as round, orange, sweet, food, living, etc. Tennis balls are round, fuzzy, yellow or green (or another color). It is not necessary for them to fill in all the lines on the chart.

❸ Next, have the students determine some overall categories into which the objects can be placed. For example, marbles, cotton balls, and oranges are round, so "Round" could be a category. Basketballs, baseballs, and footballs are all balls, so another category could be "Types of Balls." Have the students write a category at the top of each column using a PENCIL so they are able to change the categories as more items are being written down.

❹ Students will list objects in the category that describes them according to their characteristics. Some items may fit into more than one category. Basketballs can fit into both the category "Round" and the category "Types of Balls." In the chart provided, have the students write down each item in all of the categories where it fits.

❺ Have the students look at each category separately and then choose three categories to further divide into subcategories. Guide them in thinking about what the subcategories might be, trying to choose categories that allow all of the items to ultimately be listed. If necessary, they can rename some of the main categories to better fit the items listed. The names of the categories and subcategories can be adjusted as needed so that each item is listed in a category and subcategory, but it's possible that not all of the items can be placed in a category and a subcategory. This can be quite challenging. The point of this exercise is to illustrate the difficulty of trying to find a suitable organizational scheme for things with different characteristics.

## III. Conclusions

Have the students review the results they recorded for the experiment. Help them write valid conclusions based on the data they have collected. For example:

- Both oranges and cotton balls are round.
- Both cotton balls and marshmallows are white.
- Tennis balls and cotton balls are both fuzzy.

Examples of conclusions that are not valid:

- Both cotton balls and marshmallows are white. Marshmallows are sweet so cotton balls are sweet.
- Tennis balls and cotton balls are both fuzzy. Tennis balls are bouncy so cotton balls must be bouncy.

It is important to use only the data that has been collected and not make statements about the items that are not backed up by the data. It is obvious that marshmallows and cotton balls are both white, but it is not true that cotton balls are sweet. Because two or more items have one or two things in common does not mean that all things are common between them. Discuss this observation with the students.

Discuss the difference between valid and invalid conclusions. A valid conclusion is a statement that generalizes the results of the experiment, but draws only from the data collected. It does not go beyond the results of the data to include things that haven't been observed and does not connect results that should not be connected. An invalid conclusion is a statement that has not been proven by the data, or a statement that connects the data in ways that are not valid. The example given is that marshmallows are sweet and white, but although cotton balls are also white, it is invalid to say they are sweet like marshmallows.

## IV. Why?

Read this section of the *Laboratory Notebook* with your students.
Discuss any questions that might come up.

## V. Just For Fun

Students are to list 15 or more living things that can be seen without a microscope or magnifying glass. Then they will create their own taxonomic system to categorize them. There are no "right" answers.

Have them record their chart. They may want to use more paper.

# Experiment 2

# Using Agar Plates

**Materials Needed**

- plastic petri dishes[1]
- dehydrated agar powder[2]
- distilled water
- K-12 safe *E. coli* bacterial culture[3]
- inoculation loop[4]
- candle or gas flame
- cooking pot
- mixing spoon
- oven mitt or pot holder
- measuring spoons
- measuring cup
- black permanent marker
- red marker
- rubber gloves, 2 pairs

Materials available from Home Science Tools
https://www.homesciencetools.com/

1. A stack of 20 petri dishes: http://www.hometrainingtools.com/petri-dishes-plastic-20-pk/p/BE-PETRI20/

2. Agar: http://www.hometrainingtools.com/nutrient-agar-8-g-dehydrated/p/CH-AGARN08/

3. *E. coli:* http://www.hometrainingtools.com/escherichia-coli-bacteria/p/LD-ESCHCOL/

4. Inoculation loop: http://www.hometrainingtools.com/inoculating-needle-looped-end/p/BE-INOCUL/

(Product availability or item numbers may change.)

## Objectives

In this experiment students will explore how to prepare and use agar plates.

The objectives of this lesson are for students to:

- Practice using microbiology and genetics lab equipment.
- Explore how agar plates may differ, how to observe bubbles in the agar, and what happens when cultures are spread vs. streaked.

## Experiment

### I. Think About It

Read this section of the *Laboratory Notebook* with your students.

Ask questions such as the following to guide open inquiry.

> - *Do you think one agar plate can differ from another? Why or why not?*
> - *Do you think agar plates can differ in quality? Why or why not? How?*
> - *Do you think the quality of an agar plate matters? Why or why not?*
> - *Do you think bubbles or other defects in an agar plate might cause problems for bacterial growth? Why or why not?*
> - *Do you think water condensing on an agar plate might create problems for bacterial growth? Why or why not?*
> - *What do you think will happen if an agar plate dries out? Will bacteria grow? Why or why not?*

### II. Experiment 2: Using Agar Plates

Have the students read the entire experiment before writing an objective and a hypothesis.

**Objective:** Have the students think of an objective for this experiment. (What will they be learning?)

**Hypothesis:** Have the students write a hypothesis. The hypothesis can restate the objective in a statement that can be proved or disproved by their experiment. Some examples include:

> - *Bubbles will not affect bacterial growth on an agar plate.*
> - *Bubbles will make it hard to detect bacterial growth.*
> - *Water condensation will damage the surface of an agar plate.*
> - *Bacteria will not grow on a dry agar plate.*
> - *Bacteria will grow the same way on all the plates.*

## EXPERIMENT

### Part I: Preparing Agar Plates

Help the students assemble the materials for making agar plates. Keep the experimental area as clean as possible to avoid contamination.

❶ Have the students prepare a clean, flat surface on which to pour the plates. Have them spread out 18-20 petri dishes to prepare for both the *Using Agar Plates* experiment and the *Just For Fun* experiment.

❷ Have the students add 10 ml (2 teaspoons) dehydrated agar powder to 200 ml (about 1 cup) room temperature distilled water in a cooking pot and bring to a boil while stirring.

❸ The hot agar can be cooled slightly before pouring but should not cool so much that it starts to harden. Have the students use an oven mitt or pot holder while picking up the pot of agar. To prevent contamination, have them slide the lid partially off a petri dish just before they fill it and then re-cover it immediately after it is filled. Have them carefully pour a small amount of the hot agar into each of the petri dishes. It is easy to pour too much agar in one petri dish and too little in another, but they should try to put just enough agar in each petri dish to cover the bottom.

Students will need to have one or more plates that have bubbles in the agar. Each plate that contains bubbles should be marked with a red dot on the lid. If no bubbles are forming on the plates, have the students shake the pot of hot agar before pouring the last few plates.

❹ Once all the plates have been poured, they should be allowed to cool until the agar forms a hard surface.

❺ Have the students mark the lids of four plates as follows:

Take one of the plates that has bubbles and turn it upside down, agar side up. Mark it "Bubbles" and store it agar side up in the refrigerator.

Take a plate without bubbles and turn it upside down, agar side up. Mark it "Up" and store it agar side up in the refrigerator.

Take a plate without bubbles and leave it agar side down. Mark it "Down" and store it agar side down in the refrigerator.

Take a plate with or without bubbles, remove the lid, and allow it to dry until the edges of the agar start to pull away from the sides of the petri dish. This may take a few days depending on the humidity in your region. Replace the lid and mark it "Dry." Store right side up at room temperature.

❻ Have the students store the remaining plates upside down (agar side up) in the refrigerator.

## Part II: Streaking the *E. coli* Culture

Have the students wear rubber gloves during the experiment. Explain that scientists in a laboratory wear rubber gloves to prevent contamination of samples.

❶-❷ Have the students sterilize the inoculating loop by heating it in a gas flame or in a candle flame until the wire turns orange and then let it cool without touching anything. [Note: If using a candle flame, the loop should be kept close to the blue part of flame to avoid soot.]

❸ Have the students dip the loop into the tube of *E. coli* culture. They need to be careful not to let the loop touch the sides of the tube as they insert and remove it. Repeat for each plate.

❹ Have the students streak the marked plates with the inoculation loop in a zigzag pattern from one side of the petri dish to the other. The inoculated plates will then be stored agar side up at room temperature.

❺ Using one agar plate that has no bubbles and was stored in the refrigerator agar side up, the students are to pour a small volume of *E. coli* culture directly on the agar (enough to cover the agar) and then allow the agar to absorb the culture. This plate will be marked "Spread," then stored agar side up at room temperature with the other plates.

### Results

Have the students examine the plates carefully after they begin to observe growth (about 1-3 days). Have them record their observations in the chart provided.

## III. Conclusions

Have the students draw conclusions based on their observations and research. What differences did they see between the plates that were prepared in different ways?

## IV. Why?

Discuss why preparing good agar plates is important for performing good microbiological and genetics experiments. Bacterial growth may not occur properly if the agar plates have bubbles, condensation, or are dried out.

## V. Just For Fun

Have students practice different streaking patterns and then, in about 1-3 days, observe how well or poorly these streaks result in the production of single colonies.

# Experiment 3

# Using a Light Microscope

**Materials Needed**

- microscope with 4X, 10X, and 40X objective lenses. A 100X objective lens is recommended but not required. (See following pages.)
- glass microscope slides[1]
- glass microscope cover slips[2]
- immersion oil (if using 100X objective lens)[3]
- Samples:
  piece of paper with lettering
  strands of hair
  droplet of blood
  insect wing

Suggested source:
http://www.hometrainingtools.com/

1. glass microscope slides
   MS-SLIDP72 or MS-SLIDEPL

2. glass microscope cover slip
   MS-SLIDCV

3. immersion oil
   MI-IMMOIL

(At the time of this writing. Product availability or item numbers may change.)

## How to Buy a Microscope

### What to Look For

- A metal mechanical stage.
- A metal body painted with a resistant finish.
- DIN Achromatic Glass objective lenses at 4X, 10X, 40X (a 100X lens is optional but recommended).
- A focusable condenser (lens that focuses the light on the sample).
- Metal gears and screws with ball bearings for movable parts.
- Monocular (single tube) "wide field" ocular lens.
- Fluorescent lighting with an iris diaphragm.

### Price Range

$50-$150: Not recommended: These microscopes do not have the best construction or parts and are often made of plastic. These microscopes will cause frustration, discouraging students.

$150-$350: A good quality standard student microscope can be found in this price range. We recommend Great Scopes for a solid student microscope with the best parts and optics in this price range. http://www.greatscopes.com

Above $350: There are many higher end microscopes that can be purchased, but for most students these are too much microscope for their needs. However, if you have a child who is really interested in microscopy, wants to enter the medical or scientific profession, or may become a serious hobbyist, a higher end microscope would be a valuable asset.

### Objective lenses: Magnification/Resolution/Field of View/Focal Length

The objective lenses are the most important parts of the microscope. An objective lens not only magnifies the sample, but also determines the resolution. However, higher powered objective lenses with better resolution have a smaller field of view and a shorter focal length.

The resolution and working distance (focal length) of a lens is determined by its numerical aperture (NA). Following is a list of magnifications, numerical aperture, and working distance for some common achromatic objective lenses.

| Magnification | Numerical Aperture | Working Distance (mm) |
|---|---|---|
| 4X | 0.10 | 30.00 |
| 10X | 0.25 | 6.10 |
| 20X | 0.40 | 2.10 |
| 40X | 0.65 | 0.65 |
| 60X | 0.80 | 0.30 |
| 100X (oil) | 1.25 | 0.18 |

You can see as the magnification increases the numerical aperture increases (which means the resolution increases) and the working distance decreases.

Choosing the right lens for the right sample is part of the art of microscopy.

Most student projects can be achieved with a 40X objective, however a 100X objective lens can be added to make observing bacteria and small cell structures possible.

Below is a general chart showing the recommended objective lens to use for different types of samples.

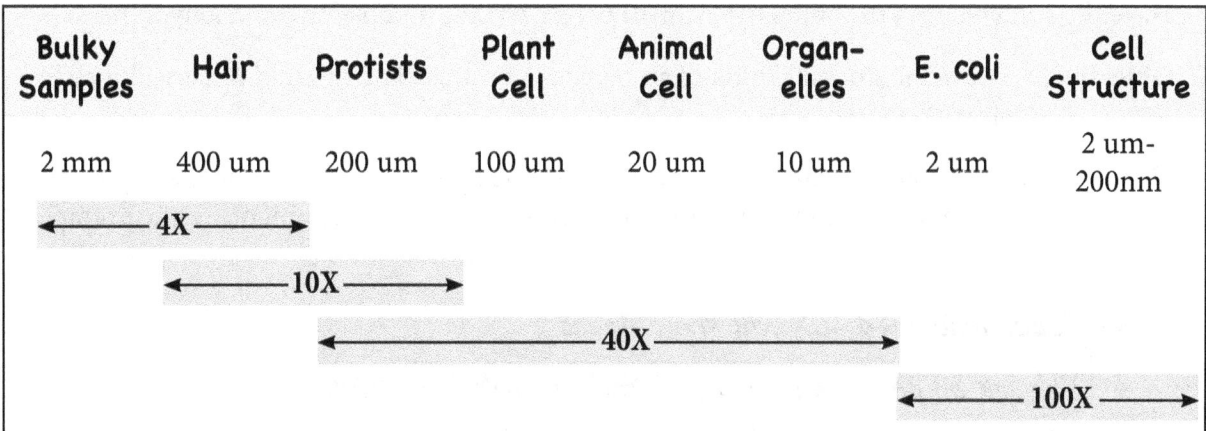

In this experiment students will explore how to use a microscope.

## Objectives

The objectives of this lesson are for students to:

- Practice using a microscope.
- Observe small details.

## Experiment

### I. Think About It

Read this section of the *Laboratory Notebook* with your students.

Ask questions such as the following to guide open inquiry.

- *Do you think a 10X lens will magnify more than a 4X? Why or why not?*
- *Do you think a 100X lens will magnify more than a 10X? Why or why not?*
- *Do you think it will be easier or harder to focus a 4X objective than a 10X objective? Why or why not?*
- *Do you think it will be easier or harder to focus a 10X objective than a 100X objective? Why or why not?*

### II. Experiment 3: Using a Light Microscope

Have the students read the entire experiment before writing an objective and a hypothesis.

**Objective:** Have the students think of an objective for this experiment. (What will they be learning?)

**Hypothesis:** Have the students write a hypothesis. The hypothesis can restate the objective in a statement that can be proved or disproved by their experiment. Some examples:

- *Paper magnified 40X will show fibers.*
- *The ink on paper will look different at 40X and 100X.*
- *I will be able to see blood cells at 40X.*
- *I will only be able to see blood cells at 100X.*

## EXPERIMENT

**NOTE:** As students are turning the turret to change lenses, help them be extremely careful not to bang the lenses on the stage or glass slide. This can damage the lenses.

### Part I: The Microscope

❶-❽ Have the students follow the instructions in the *Laboratory Notebook* to observe the parts of the microscope and how the different parts work. Have them label the parts of the microscope in the diagram provided.

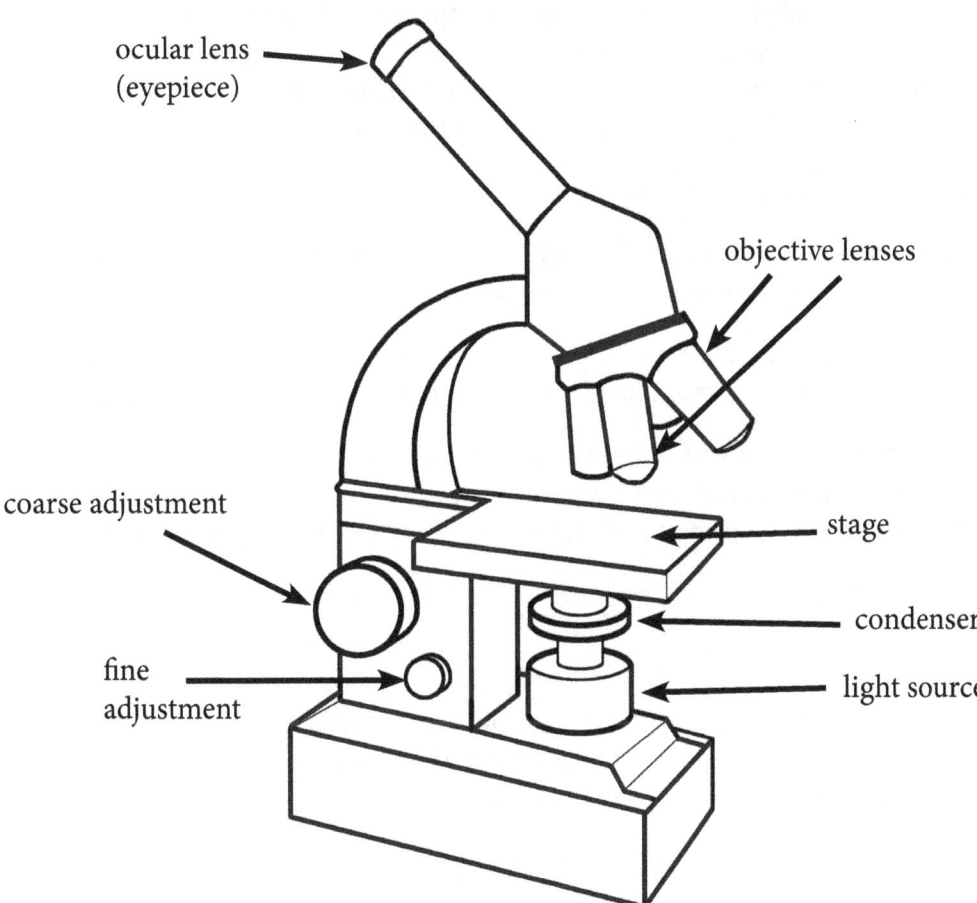

### Part II: Observing a Sample

Help the students avoid scraping any of the samples with the objective lenses.

1-6. Have the students take a small piece of paper that has lettering on it and place it on a glass slide in the microscope without a cover slip. Have them examine it using the 4X and 10X objective as instructed. Paper is considered a "bulky" sample, so the low magnification lenses are used. Do not have the students use a higher power lens because it could get damaged. In the spaces provided, have them record their observations.

7. Have the students turn the turret to move the lowest power objective back into place. **[NOTE: If they have a 100X objective, do not let them rotate the turret through this lens. If the lens scrapes the slide, it can ruin the lens.** Instead, have them turn the turret in the opposite direction until the lowest power lens is back in place]

8-14. Have the students create a glass slide with fresh blood. Have the them wash their hands and then collect a drop of blood by pricking a finger with a needle that has been sterilized in a candle flame. Blood is a great sample to observe in a microscope because it will flow for a few minutes before it dries which will coat the area under the coverslip, making it easy to find and focus. Have the students start with the lowest power objective lens (4X) and move to the higher powered objective lenses one at a time, focusing each one as they go. If you plan to have them use a 100X objective lens, help them place a drop of immersion oil on the coverslip and very carefully rotate the turret to click the lens into place. It is very easy to smash the sample surface and ruin the lens, and if this seems about to happen, have them back the lens up one or two turns with the fine adjust knob. Have them always use only the fine adjustment knob with the 100X lens to avoid hitting the slide.

Have them adjust the condenser to get more light.

Each time the students look at the sample through a different lens, have them record their observations in writing and by drawing what they see.

15. Have the students rotate the turret to move the 100X immersion lens away from the sample. Then they can use the coarse adjustment knob to lower the stage and remove the sample.

16. Have them repeat the experiment with other samples, such as hair, pond water, or an insect wing. Have them look at each sample beginning with the lowest power lens and rotating through to the highest power, focusing the image each time before moving to the next highest power lens.

## III. Conclusions

Discuss how easy or difficult the students found the use of the microscope. Using a microscope is an art, and learning how to use one correctly takes time and patience. Have the students note whether their conclusions support or do not support their hypothesis.

## IV. Why?

Read this section of the *Laboratory Notebook* with your students.
Discuss how the different objectives magnified the samples with different degrees of resolution and focal length.

## V. Just For Fun

Have the students use the microscope to observe other samples of their choice. Have them write and draw their observations.

# Experiment 4

# What's in Spit?

**Materials Needed**

- tincture of iodine [**VERY POISONOUS—DO NOT ALLOW STUDENTS TO EAT** any food items that have iodine on them]
- bread, 1 slice
- timer
- wax paper
- marking pen
- cup
- refrigerator
- a green vegetable
- one or more other vegetables or fruits

## Objectives

In this experiment students will observe a chemical reaction in part of the metabolic process.

The objectives of this lesson are for students to:

- Observe evidence of a chemical reactions that happens in the body.
- Observe how the process of digestion of food begins.

## Experiment

### I. Think About It

Read this section of the *Laboratory Notebook* with your students.

Ask questions such as the following to guide open inquiry.

> - *How do you think your body digests food?*
> - *Why do you think food needs to be chewed?*
> - *Why do you think you have saliva in your mouth?*
> - *Do you think your body could digest food if you didn't have saliva? Why or why not?*
> - *Do you think digestion requires chemical reactions? Why or why not?*
> - *Do you think chemical reactions happen in your mouth? Why or why not?*

### II. Experiment 4: What's in Spit?

In this experiment students will investigate the part of the digestive process carried out by proteins in saliva. Have the students read the experiment before writing an objective and hypothesis.

### EXPERIMENT

An example **Objective:** *We will investigate what saliva does to bread.*
An example **Hypothesis:** *We will be able to test changes in the bread by using iodine.*

❶ Have the students break the bread into several small (bite size) pieces.

❷ Students will chew one piece of bread for 30 seconds, another piece for 1 minute, and a third for several minutes. Have them set a timer for each.

❸ After each chewing time is up, have the students spit the chewed bread onto a piece of wax paper and use a marker to record the length of time it was chewed.

❹ Have the students place one small piece of unchewed bread next to each piece of chewed bread.

❺ Have the students put a drop of iodine on each of the pieces of bread, both chewed and unchewed.

❻ Have them record their results in the chart provided. They should observe that the color resulting from the iodine reacting with the bread that has been chewed for longer times is not as black as with unchewed bread or bread that has not been chewed as much.

❼ Have the students collect saliva by spitting into a cup several times. Then they will take two small pieces of bread and soak both in the saliva. They can add more saliva to the cup, if needed.

❽ Have the students place each piece of soaked bread on a separate piece of wax paper and put one in the refrigerator and leave one out at room temperature.

❾ After 30 minutes, have the students test each piece of bread by putting a drop of iodine on each.

❿ Have them record their results. The refrigerated bread should turn more black than the unrefrigerated bread because the cold temperature slows the chemical reaction.

## III. Conclusions

Have the students review the results they recorded for the experiment. Have them draw conclusions based on the data they collected.

## IV. Why?

Read this section of the *Laboratory Notebook* with your students.
Discuss any questions that might come up.

## V. Just For Fun

Have the students repeat the experiment with celery, kale, or another green vegetable. The iodine should not change color. Have them test one or more other vegetables or fruits. Following are some expected results, but any vegetables or fruits can be tested:

Color change:     apple, banana, pasta, potato, yam

No color change:   celery, kale, spinach, green bell pepper

Have the students draw conclusions from their results.

# Experiment 5
## Inside the Cell

**Materials Needed**
- pencil
- colored pencils or crayons
- student-selected materials to build a model cell

## Objectives

In this experiment students will explore cells and their structure.

The objectives of this lesson are for students to:

- Observe that cells are highly complex and highly ordered.
- Compare the features of three different types of cells, observing similarities and differences.

## Experiment

### I. Think About It

Read this section of the *Laboratory Notebook* with your students.

Ask questions such as the following to guide open inquiry.

- *Why do you think scientists study cells?*
- *Do you think it is helpful for scientists to put different kinds of cells into groups? Why or why not?*
- *Do you think it's important that there are different kinds of cells? Why or why not?*
- *What do you think life would be like if all cells were exactly the same?*
- *How do you think you could tell one type of cell from another?*

### II. Experiment 5: Inside the Cell

In this exercise students will examine the similarities and differences between three cell types.

All cells share some common features. One such feature is DNA (deoxyribonucleic acid). DNA is often referred to as the genetic code. Almost every cell has DNA. The DNA in a cell contains many volumes of information that the cell needs to make proteins, metabolize nutrients, grow, and divide.

Another feature common to cells is that they have ribosomes which make proteins from RNA (ribonucleic acid). RNA is different from DNA but is still a nucleic acid. RNA is made from DNA and proteins are made from RNA.

DNA->RNA->proteins

In living cells there are no known exceptions to this paradigm. Proteins are always made from RNA, and the RNA used to make proteins is always made from DNA.

Have the students read the entire experiment.

**Objective:** An objective is provided.

## EXPERIMENT

The first part of the experiment has questions for students to answer. Following are examples of possible answers. Answers may vary.

- *List some things you observe in the drawings in the textbook that are similar for all three cell types:*

    (Examples. Answers may vary.)
    All cells contain DNA.
    All cells contain ribosomes.
    All cells have something that holds them together, like a cell wall or plasma membrane.

- *List some observations of things that are different:*

    (Examples. Answers may vary.)
    Bacterial prokaryotic cells do not have a nucleus.
    Animal cells do not have a cell wall.
    Plant cells contain chloroplasts, but animal cells do not.

- *List the function of each of the following:*

    nucleus    In eukaryotic cells, the nucleus holds together the DNA and the proteins needed to use the DNA.

    mitochondria    organelles that make energy; found in plant and animal cells

    chloroplasts    organelles that use the Sun's energy to make food; found in plant cells

    cell wall    stiff outer membrane found in plant cells; makes the plant sturdy

    lysosome    the place where big molecules get broken down

    peroxisome    the place where poisons in the cell are removed

# Experiment 5: Inside the Cell

❶ Students are asked to list differences between bacteria, plants, and animals and why their cells may need to be different. There is a chart provided to be filled in. Answers will vary and there are no "right" answers.

To learn more, students can do research online or at the library.

Following are some facts about bacteria:

- They can be spherical, rod shaped, or spiral.
- They live in many different environments, including soil, water, organic matter, and in the bodies of plants and animals.
- They make their own food (are autotrophic), or live on decaying matter (are saprophytic), or live off a live host (are parasitic).
- They can be either beneficial or harmful to humans.

Some facts about plants and animals:

Plants have organs, as do animals, and therefore need to be made of many different types of cells. Have the students think about the different parts of a plant, such as the leaves and roots, and discuss what the cells of each part might need to do. (For example: Root cells need to take up minerals from the soil. Since roots are in the dirt, they do not have to be green like leaves. The leaves are green because they need to use chloroplasts for collecting light.)

Discuss how plant cells differ from animal cells. For example, plants don't have bones, and they don't usually move, so they don't need muscles like some animals. Have the students think of a variety of animals, such as deer, fish, and frogs, and then discuss the differences between them. Next, have them write down why there are different types of cells in these different creatures.

❷ Bacterial prokaryotic cell  ❸ Animal cell  ❹ Plant cell

Have the students fill in the blanks in the drawings on these pages. Have them first identify what type of cell they are looking at. Have them label as much of the drawings as they can without looking at the *Student Textbook* and then refer to the textbook to finish the labeling. When they have filled in all the labels, have them color the different parts of the cell. The colors do not need to match those in the textbook.

As the students fill in the blanks, discuss the functions of the various parts of the cell. Point out how the structures differ and, where possible, point out how the structure of the part matches its function. For example, the flagellum looks like a whip and is used in a whipping motion for swimming. The cell membrane and cell wall are used to enclose the contents of the cell, are thin, and extend around the outside of the cell.

## III. Conclusions

Have the students arrive at some conclusions about cells based on what they have learned in this chapter.

Some examples include:

> * *All living things are made of cells.*
> * *All living things have DNA.*
> * *Not all cells are alike.*

Answers may vary.

## IV. Why?

Read this section of the *Laboratory Notebook* with your students.
Discuss any questions that might come up.

## V. Just For Fun

Students will make a simple model of a cell. They can select a few features to model rather than trying to include them all.

Have them decide which of the three cells they would like to model and then make a list of materials they think they could use to build their model. They can use one type of material, such as colored clay, or a combination of materials, such as clay, paperclips, string, wire, food items, etc. Encourage them to use their imagination in selecting materials to represent the different cell features.

# Experiment 6

# Wash Your Hands!

**Materials Needed**

- dehydrated agar powder*
- distilled water
- cooking pot
- measuring spoons
- measuring cup
- cup
- plastic petri dishes (20)**
- cotton swabs
- permanent marker
- oven mitt or pot holder

Items available from Home Science Tools
https://www.homesciencetools.com/
(Product availability or item numbers may change.)

* Agar:
  http://www.hometrainingtools.com/nutrient-agar-8-g-dehydrated/p/CH-AGARN08/

** Stack of 20 petri dishes:
  http://www.hometrainingtools.com/petri-dishes-plastic-20-pk/p/BE-PETRI20/

## Objectives

In this experiment students will grow bacterial cultures and practice using controls.

The objectives of this lesson are for the students to:

- Explore how the use of control experiments is required to verify data.
- Observe how scientific explanations emphasize evidence and use scientific principles.

## Experiment

### I. Think About It

Read this section of the *Laboratory Notebook* with your students.

Discuss how bacteria and viruses are small organisms that we can't see with our eyes alone and that they can live on surfaces and inside the body.

Discuss how some bacteria are healthful and help us digest our food and some bacteria are harmful and make us sick.

Explain that scientists use agar plates to grow bacteria. Agar is made from various types of seaweed and is used as a thickening agent. The dehydrated agar used in this experiment is called nutrient agar and contains not only agar but also vitamins, amino acids, carbon, and nitrogen derived from beef extract. Nutrient agar will give off a slight odor when cooked.

Ask questions such as the following to guide open inquiry.

- *Do you think bacteria are important? Why or why not?*
- *Do you think all bacteria are harmful? Why or why not?*
- *Do you think you can ever see bacteria? Why or why not?*
- *Why do you need to wash your hands after being outside?*
- *Why do you need to wash your hands after using the bathroom?*
- *Do you think your hands are "clean" after you wash them? Why or why not?*
- *Do you think the surfaces in your house, like the computer keyboard and kitchen doorknob might have bacteria on them? Why or why not?*

## II. Experiment 6: Wash Your Hands!

Have the students read the entire experiment.

**Objective:** Have the students write an objective.
**Hypothesis:** Have the students write a hypothesis.

### EXPERIMENT

### Part I: Preparing Agar Plates

Help the students assemble the materials for making agar plates. Making agar plates is not difficult but may take a little practice. Keep the experimental area as clean as possible to avoid contamination.

❶ Have the students prepare a clean, flat surface on which to pour the plates. Have them spread out 18-20 petri dishes to use for both the *Wash Your Hands* experiment and the *Just For Fun* experiment.

❷ Have the students follow the directions for preparing the agar.

❸ The hot agar can be cooled slightly before pouring but should not cool so much that it starts to harden. Have the students use an oven mitt or pot holder while picking up the pot of agar. To prevent contamination, have them slide the lid partially off a petri dish just before they fill it and then re-cover it immediately after it is filled. Have them carefully pour a small amount of the hot agar into each of the petri dishes. It is easy to pour too much agar in one petri dish and too little in another, but they should try to put just enough agar in each petri dish to cover the bottom.

❹-❺ Once the agar has solidified, the petri dishes can be stored upside down with the agar on top. This keeps condensation from collecting on the surface of the agar. Have the students stack the inverted petri dishes and place them in the refrigerator.

### Part II: Testing for Bacteria

❶-❷ The students will prepare two petri dishes to act as controls for the experiment. The control plates will indicate if their agar or their water is contaminated. If the agar control shows growth after a few days, this means the agar was contaminated during preparation. The experiment can be repeated. Be careful to use a clean area. If the water control shows growth after a few days, the water can be boiled and the experiment repeated. Most of the time, both the agar and the water controls will be clean. Bacterial growth occurring after several weeks is normal.

Have the students take one petri dish from the refrigerator, label it "Agar," and return it to the refrigerator. A second petri dish will be labeled 'Water" and left out at room temperature.

❸ Students will prepare the control test for the water being used. Have them swirl a cotton swab in the distilled water and shake off the excess water. Then have them "streak" the plate labeled "Water" by gently moving the swab in a zigzag motion across the agar from one side of the petri dish to the other, being careful not to break the hardened surface of the agar.

❹-❺ They will now test their hands for the presence of bacteria. Have them remove another petri dish from the refrigerator and label it "Hands." Have them take an unused cotton swab, swirl it in the distilled water, and shake off the excess Then they will rub the swab on their fingertips and streak the agar plate as in Step ❸.

❻ Have the students choose some surfaces they'd like to test and write the names of the surfaces in the chart provided.

❼ Have the students remove the petri dishes one at a time from the refrigerator as they test each surface they've chosen. The plate labeled "Agar" should be left in the refrigerator. Have the students refer to their chart, and keeping the petri dish agar side up, label it with the name of the surface to be tested. The streaked plates will be left out at room temperature.

❽ Have the students prepare each plate using the same method as before.

❾ Once all the plates have been prepared, have the students put them in a stack agar side up. Have them remove the "Agar" plate from the refrigerator and add it to the stack. The plates will be left to incubate for a week to ten days. They should not be put in a bag.

**Results**

After 7-10 days, have the students examine each of their plates and record their observations in the table provided.

### III. Conclusions

Have the students review the results they recorded for the experiment. Have them note whether there was bacterial growth on the control plates. Then have them compare their control plates to their test plates to look for bacterial growth. Have them draw conclusions based on their data—what they actually observed and not what they think should have happened.

### IV. Why?

Read this section of the *Laboratory Notebook* with your students.
Discuss any questions that might come up.

### V. Just For Fun

Have the students retest the surfaces that showed bacterial growth in Part II of the experiment. This time they will wash each surface with a different household cleaner and repeat the experiment to see if they can determine how effective the different cleaners are at removing bacteria. A table is provided for them to record their results.

# Experiment 7

# Observing Protists

**Materials Needed**

- microscope with a 10X objective
- microscope depression slides [1]
- 10 or more eyedroppers
- fresh pond water or water mixed with soil
- protozoa study kit [2] (must be used within 1-2 days of arrival)
- methyl cellulose [3]
- measuring cup and measuring spoons
- baker's yeast
- distilled water
- Eosin Y stain [4]

As of this writing, the following materials are available from Home Science Tools: www.hometrainingtools.com:

1. Glass Depression Slides, MS-SLIDC72 or MS-SLIDC12
2. Basic Protozoa Set, LD-PROBASC
3. Methyl Cellulose, CH-METHCEL
4. Eosin Y, CH-EOSIN

(Product availability or item numbers may change.)

## Objectives

In this experiment students will be introduced to the microscopic organisms known as protists.

The objectives of this lesson are for students to:

- Observe how protists move and eat.
- Use a microscope to make observations.

## Experiment

### I. Think About It

Read this section of the *Laboratory Notebook* with your students.

Ask questions such as the following to guide open inquiry.

- *How many different ways do you think protists move?*
- *Do you think a paramecium moves more like a euglena or an amoeba? Why?*
- *How many different methods can you think of that protists use to eat?*
- *Do you think using cilia, a flagellum, or pseudopodia is the most efficient way for a protist to move? Why?*
- *Do you think it is easier for a paramecium to get around than for an amoeba? Why or why not?*

### II. Experiment 7: Observing Protists—How Do They Move?

In this experiment students will examine the three different types of protists discussed in this chapter of the *Student Textbook*. They will then examine pond water or water mixed with soil to identify individual protists based on their method of movement.

It may take some time for younger students to align their eye directly into the lens so that the sample is visible. Also, viewing tiny organisms through the small eyepiece of a microscope can be difficult and requires some patience. These organisms can swim rapidly through the field of view, and it is easy to get frustrated trying to observe them. Methyl cellulose will help slow the organisms down without killing them. Patience with this experiment is a must.

Because students will be using slides with a concavity for the sample, they will not need to use cover slips.

Have the students read the entire experiment before writing an objective and a hypothesis.

**Objective:** Have the students write an objective. For example:

- *In this experiment, three types of protists will be observed. We will see how they move in different ways.*

**Hypothesis:** Have the students write a hypothesis. Some examples:

- *We can tell the difference between ciliates, flagellates, and amoebae.*
- *We can tell the difference between ciliates, flagellates, and amoebae in pond water by how they move.*

## EXPERIMENT—Part A

Have the students set up the microscope.

❶ Have the students position a slide in the microscope and use an eyedropper to put a droplet of one of the protist samples on the slide.

❷ Have the students observe how the protists move. A droplet of methyl cellulose can be added to the protist sample on the slide to slow the movement of the protists.

A euglena will tend to move in a single direction, or it may not move at all but "hover" just under the light.

A paramecium will move all over the place. It will roll, move forward and backward, and spin. There are usually other things in the water with the paramecium. Have the students note what happens when the paramecium "bumps" into other objects or other paramecia.

The amoebae move very slowly, and it can be difficult to observe them. They are usually on the bottom of the container. Allow the container to sit for 30 minutes, and then have the students remove some solution from the very bottom, placing it on a slide. The amoebae should be visible but may be difficult to see since they are clear.

❸ Have the students draw the protist they are observing and write their observations. Boxes are provided in the *Results* section.

❹ Have the students repeat Steps ❶-❸ using the remaining two protist samples. Have them use a new eyedropper for each sample.

## Results—Part A

Boxes are provided for students to record their results.

## EXPERIMENT—Part B

Have the students repeat the experiment, this time looking for protists in pond water or water mixed with soil. Have them use a new eyedropper to place a droplet of fresh pond water (or water mixed with soil) on a slide in the microscope. Have them look for protists and try to determine the types of protists they observe based on how the organisms move. Have the students refer to the notes they made in *Part A* for comparison. Space is provided in the *Results — Part B* section for writing and drawing their observations. Have them record information for as many different organisms as they can find.

## III. Conclusions

Have the students review the results they recorded for the experiment. Have them draw conclusions based on the data they collected. If the experiment did not work, this should be written as a conclusion.

## IV. Why?

Read this section of the *Laboratory Notebook* with your students.
Discuss any questions that might come up.

## V. Just For Fun: How Do They Eat?

Students will perform another experiment to observe how two different protists eat. Eosin Y stained baker's yeast will be used as food to be ingested by the protists. It may take some time for this observation. Once ingested by a protozoan, the red stained yeast will turn blue.

Have the students read the entire experiment. Have the students predict whether or not the protists will eat the yeast. Then have them write the objective and hypothesis.

❶ Have the students follow the directions to color the yeast with Eosin Y stain:

Add 5 milliliters (one teaspoon) of dried yeast to 120 milliliters (1/2 cup) of distilled water. Allow it to dissolve.

Add one droplet of Eosin Y stain to one droplet of yeast mixture. Look at the mixture under the microscope. You should be able to see individual yeast cells that are stained red.

❷ Have the students place a droplet of the amoeba sample on a glass slide that has been correctly positioned in the microscope. (For the amoebae, remind the students to gather the sample from the bottom of the container it comes in.)

❸ Have the students add a small droplet of the yeast stained with Eosin Y to the droplet of protist solution that is on the slide.

❹ Have the students observe the protists through the microscope, noting the red-colored yeast. Have them describe how the protist eats, writing down as many observations as they can and drawing one of the protists eating. It may take some patience to find protists eating.

❺ Have the students repeat steps 2-4, this time using the paramecium sample.

# Experiment 8

## Moldy Growth

### Materials Needed

- dehydrated agar powder*
- distilled water
- cooking pot
- measuring spoons
- measuring cup
- plastic petri dishes**
- permanent marker
- oven mitt or pot holder
- jar with lid (big enough to hold 235 ml (about 1 cup) liquid
- 1 slice of bread, preferably preservative free
- small clear plastic bag
- white vinegar
- bleach
- borax
- mold or mildew cleaner
- 1-2 pairs rubber gloves

Items available from Home Science Tools
https://www.homesciencetools.com/
(Product availability or item #s may change.)

* Agar: http://www.hometrainingtools.com/nutrient-agar-8-g-dehydrated/p/CH-AGARN08/

** 20 petri dishes: http://www.hometrainingtools.com/petri-dishes-plastic-20-pk/p/BE-PETRI20/

## Objectives

In this experiment students will gain more experience in making agar plates and using controls.

The objectives of this lesson are for students to:

- Practice the lab technique of pouring and using agar plates.
- Observe mold growth and products that prevent the growth of mold.

## Experiment

### I. Think About It

Read this section of the *Laboratory Notebook* with your students.

Ask questions such as the following to guide open inquiry.

- *How do you think mold and bacteria are different?*
- *Do you think products that kill bacteria can kill mold? Why or why not?*
- *Do you think it is easier or harder to prevent mold from growing in wet, warm, and dark areas? Why or why not?*
- *Do you think it is easier or harder to prevent mold from growing in dry, cold, and light areas? Why or why not?*

### II. Experiment 8: Moldy Growth

Have the students read the entire experiment before writing an objective and a hypothesis.

**Objective:** Have the students think of an objective for this experiment. (What will they be learning?)

**Hypothesis:** Have the students write a hypothesis. The hypothesis can restate the objective in a statement that can be proved or disproved by their experiment. Some examples include:

- *Bleach will kill mold.*
- *Bleach will not kill mold.*
- *Borax will kill mold better than bleach.*
- *Vinegar will kill mold better than bleach or borax.*

# EXPERIMENT

## Part I: Pour Agar Plates

Have the students follow the directions in *Experiment 6* for making agar plates. They will need six plates that don't have bubbles in the agar for this experiment and one or more additional bubble-free plates for the *Just For Fun* section.

Have the students store the plates upside down (agar side up) in the refrigerator until they are ready to do the experiment.

## Part II: Observing Mold

In this experiment students will observe which household products kill mold, and whether they prevent growth. The agar on the plates is a food source for the mold from the bread.

❶ Students are to make moldy bread by taking a piece of bread (preferably without preservatives), placing it in a plastic bag, adding a teaspoon of water, sealing the bag, and then letting it sit in a dark, warm area for several days.

❷ Have the students put on rubber gloves. Have them cut 5 small cubes from the moldy bread and save the rest for the *Just For Fun* experiment.

❸-❹ Have the students mark agar plates as follows: *Control, None, Vinegar, Bleach, Borax,* and *Mold Cleaner.* They will put the marked plates agar side down.

❺-❻ Students will add 5 ml (1 tsp.) of white vinegar to the *Vinegar* plate and tilt the plate gently back and forth to evenly cover the agar. Have them repeat for the *Bleach* plate.

❼ Have the students measure 235 ml (1 cup) of distilled water into a jar, add 5 ml (1 tsp.) of borax, put the lid on the jar, and shake it until the borax is completely dissolved. They will then measure 5 ml (1 tsp.) of the mixture and put it on the *Borax* plate, tilting the plate back and forth until the agar is covered evenly.

❽ Students will add 5 ml (1 tsp.) of mold or mildew cleaner to the *Mold Cleaner* plate and tilt the plate back and forth to evenly cover the agar

❾ Have the students add a cube of moldy bread to each petri dish except the one marked *Control.*

❿ Plates should be moved to a warm, dark room, optimally 27°C (80°F), and stored agar side down.

## Results

Have the students observe the plates for several days, and then in the table provided, write and draw their observations.

## III. Conclusions

Have the students review the results they recorded for the experiment. Have them draw conclusions based on the data they collected.

Have them compare their results to the control plates (*Control* and *None*) and observe which cleaners killed mold, prevented mold growth, and both killed and prevented mold growth.

## IV. Why?

Read this section of the *Laboratory Notebook* with your students.
Discuss how cleaners kill and prevent mold growth. Discuss any questions that might come up.

## V. Just For Fun

### Part I

Have the students repeat the experiment, this time covering the agar with 5 ml (1 tsp.) of 3% hydrogen peroxide solution and then adding the moldy bread. If they have additional prepared agar plates left, they can try other household products of their choice to see if they kill or prevent mold.

Have them record their results in the space provided.

### Part II

Have the students look at a small piece of the moldy bread through their microscope. Have them record their observations.

# Experiment 9
## Identifying Plants

**Materials Needed**

- colored pencils
- handheld magnifying glass
- field notebook (Have students begin a field notebook or use an existing one. If it is a new notebook, one that has blank or faintly lined pages will work best.)
- backpack with water and snacks
- 2 plant pots
- potting soil and water
- corn seeds, 8 or more with packet
- bean seeds, 8 or more with packet
- warm, sunny location

**Optional**

- field guide to the plants book
- iPad, camera, or smartphone with camera
- plant identification app
  At the time of this writing some available apps are: Like That Garden, Leafsnap (trees), PlantNet, and NatureGate. Because apps can be device specific and change over time, you and your students will need to do some online research to find the best app for this experiment.

## Objectives

In this experiment students begin learning how to identify different types of plants by observing their features.

The objectives of this lesson are for students to:

- Make and record careful observations.
- Learn how plants can be categorized by observing their features.

## Experiment

### I. Think About It

Read this section of the *Laboratory Notebook* with your students.

Ask questions such as the following to guide open inquiry.

- *What features of a rose flower tell you it's a rose?*
- *If a rose plant doesn't have a flower on it, what features would tell you it's a rose?*
- *If you wanted to put different plants into groups that had similar features, what features do you think you would use?*
- *Think of some plants that you have observed. What shape leaf does each plant have? What does the pattern of the veins look like?*
- *Think of some flowers that you like. How do you think they are the same and how are they different?*
- *Do you think it could be important to be able to identify a particular plant? Why or why not?*
- *Think of a place where you walk frequently. How many different types of plants do you think you see there? How many different individual plants do you think you see?*

### II. Experiment 9: Identifying Plants

Have the students read the entire experiment before writing an objective and a hypothesis.

**Objective:** Have the students write an objective based on what they think they will be learning.

**Hypothesis:** This is an observational experiment so there is no hypothesis.

## EXPERIMENT

In this experiment students will observe plants and their ecosystems and use a field notebook to record their observations.

❶ Help the students choose a location where they can observe a variety of plants. If you live in a city, you might arrange a trip to a local park or a hike somewhere outside the city limits.

❷ Before going out on the hike, students are asked to prepare their field notebook by writing down the plant classification information from the chart in the section *Classification of Plants* in the *Student Textbook*. They can list the common names of the phyla and skip the scientific names. The act of writing information can help students remember it. They are also asked to list the features of monocot and dicot flowers and leaves from the chart in the section *Seeded Vascular Plants*. They will probably not be able to observe the root systems or internal structure of the stems. The Plant Identification Guide at the end of the *Experiment* section in the *Laboratory Notebook* can be torn out and taken along on the hike.

❸ Have the students spend a few hours walking or hiking in the chosen location, observing the plants around them.

❹-❺ Students are to find a small area where different types of plants are growing and observe the various plants. They will then choose 3-5 plants to observe more closely. These can be any type of plant: grasses, mosses, flowering plants, trees, etc.

❻ Using a blank notebook, students can start a field notebook that they can add drawings and notes to over time. If they already have an existing field notebook, they can use this one. Drawing or sketching plants will help students observe their features more closely, noting details such as the shape of the leaves, how they are arranged, and what pattern the veins make. The drawings don't have to be realistic renderings. Students are directed to draw or sketch the whole plant first and then select a small area of the plant to examine with a magnifying glass and draw the details they see. Next to the drawing, or as labels on the drawing, students are to make written notes about some of the features observed. In addition to doing the drawings, students can also photograph the plants and later print the photos and attach them in their field notebook. They are also to observe and take notes about the ecosystem the plants live in (shady, sunny, moist, dry, sandy soil, etc.)

❼ To help identify each plant by its type, students are instructed to refer to the notes they made about plant classifications and monocot vs. dicot features and to use the Plant Identification Guide. They can also use a field guide to the plants book or a plant identification app for their mobile device. Have them first make a guess based on their observations of the plant and then look at a book or app for verification.

## Results

Results are recorded in the field notebook.

## III. Conclusions

Have the students review the results they recorded for the experiment. Have them draw conclusions about plant identification based on their observations and the data they collected. How easy or difficult is it to identify plants? Why? What did they discover about plants and ecosystems?

## IV. Why?

Read this section of the *Laboratory Notebook* with your students.
Discuss any questions that might come up.

## V. Just For Fun

### Monocot vs. Dicot

In this experiment students will have a chance to observe the differences between a corn plant, which is a monocot, and a bean plant, which is a dicot.

❶ Have the students fill two plant pots with potting soil.

❷ Corn and bean seeds generally have different germination rates. Have the students check the seed packets for the number of days the seeds will take to germinate. Based on the germination rate shown on the seed packets, help them figure out how many days apart to plant the corn and bean seeds so they are more likely to break the ground on the same day.

❸ Have the students plant 8 or more corn seeds in one pot and 8 or more bean seeds in the other pot according to their estimate in Step ❷. Have them label the pots with the type of seed, put them in a warm, sunny location, and water thoroughly. Have the students check the soil daily, add water as needed to keep the soil moist, and look for sprouts.

❹ Have them note the dates on which sprouts break the ground and how many of which type of plant. Have them record their observations in their field notebook. Did they get the results they predicted? Why or why not? Do the sprouts look the same or different?

❺ Have the students pull up one corn sprout and one bean sprout that have broken ground and observe the roots. Have them pull up another sprout of each type every few days to observe the development of the roots until they have 1 or 2 plants left that they can let grow for some time. Have them record their observations each time.

❻ Have the students observe the plants as they grow. Have them record their observations, noting how the features of each type of plant develop over time and also noting similarities and differences between the two types of plants.

❼ Students should end up with 1 to 2 plants to tend and observe.

# Experiment 10

# Take Away the Light

**Materials Needed**

- plant with at least 6 flat, green leaves (a tree may be used)
- lightweight cardboard or construction paper—enough to cut out 6 pieces that are bigger than a leaf
- scissors
- tape
- 2 small jars
- marking pen
- 4 or more plant pots
- potting soil
- bean seeds (12 or more)

## Objectives

In this experiment students will be introduced to photosynthesis.

The objectives of this lesson are for students to:

- Observe some conditions that plants require to grow and be healthy.
- Explore conditions needed for photosynthesis to occur.

## Experiment

### I. Think About It

Read this section of the *Laboratory Notebook* with your students.

Ask questions such as the following to guide open inquiry.

> - *What conditions do you think a plant requires to grow and be healthy?*
> - *How do you think a plant gets food and nutrients?*
> - *What do you think a plant does with sunlight that falls on its leaves?*
> - *Do you think a plant that has leaves could live in your closet? In the garage? In the living room? Outside? Why?*
> - *Do you think photosynthesis is important to all life on the planet or just to plants? Why?*

### II. Experiment 10: Take Away the Light

In this experiment students will examine the effects of removing light from the leaves of a photosynthetic plant. The results for this experiment may vary depending on the type of plant that is used. A tree with leaves can also be used.

Have the students first read the experiment through to determine what is being investigated. Then have them write the objective.

**Objective:** (What will they be learning?) For example:

> - *We will place plant leaves under different conditions and after several days observe whether light and water are sufficient to keep a leaf alive whether or not it is attached to the plant.*

Next, discuss the possible outcomes for each of the leaves. Ask the students questions such as the following before they write the hypothesis.

- *What do you think will happen to Leaf 1? It is uncovered and attached to the plant. It can get sunlight and also can get water and other nutrients from the rest of the plant.*

- *What do you think will happen to Leaf 2? It is covered and attached to the plant. This means it cannot get any sunlight, but it does get water and other nutrients from the rest of the plant.*

- *What will happen to Leaf 3? It is uncovered (will get sunlight) and not attached to the plant (no additional nutrients), but it is placed in water, so it will receive water.*

- *What will happen to Leaf 4? It is covered (will not get sunlight), is unattached (will not get additional nutrients), but it will get water.*

- *What will happen to Leaf 5? It is uncovered (will get sunlight) but is unattached and will get no water or nutrients.*

- *What will happen to Leaf 6? It is covered (will get no sunlight), and it will not get any water or additional nutrients.*

Have the students guess which leaf will die first, which will not die, and which may survive a short time. Have them write a suitable hypothesis based on this discussion.

**Hypothesis:** Some examples:

- *All of the leaves not in water will die.*
- *Only the leaves without sunlight will die.*
- *Only Leaf 6 will die.*
- *Only Leaf 1 will stay healthy.*

## EXPERIMENT

1. Have the students cut out 6 pieces of cardboard or construction paper that are large enough to completely cover a leaf. These will be used to cover the front and back of 3 leaves.

2. Six different leaves will be tested. Two of the leaves will be left on the plant (attached) and four leaves will be removed from the plant (unattached).

❸ Have the students tape two of the cut out pieces so they cover the front and back of one of the leaves that is attached to the plant. At least one other leaf will stay attached to the plant and remain uncovered.

❹ Have the students remove four leaves from the plant and cover two of them with cardboard. The four unattached leaves will be either: covered with cardboard and the stem placed in water, covered out of water, uncovered with the stem in water, or uncovered out of water.

❺ Have the students use a marking pen to label the leaves in the following manner:

Leaf 1: **UA** — uncovered, attached

Leaf 2: **CA** — covered, attached

Leaf 3: **UUW** — uncovered, unattached, in water

Leaf 4: **CUW** — covered, unattached, in water

Leaf 5: **UU** — uncovered, unattached (no water)

Leaf 6: **CU** — covered, unattached (no water)

❻ Have the students fill two small jars with water and place one covered unattached leaf in one jar and one uncovered unattached leaf in the second jar so that the stems stay submerged and the leaves are out of the water. Have the students check the water level every day over the course of the experiment to make sure there is enough water in the jars.

❼ After several days have the students make daily observations of the changes to the leaves. They will need to carefully remove the cardboard from the covered leaves to make their observations, and then re-tape it. Have them record their observations in the *Results* section.

## Results

A chart is provided for recording observations. Short one or two word descriptions like, "green," "green with some brown," "mostly brown," "wrinkled," are fine.

## III. Conclusions

Have the students review the results they recorded for the experiment. Have them write valid conclusions based on the data they collected.

For example:

- Leaf 1 survived for one week.
- Leaf 2 did not survive past two days.
- Leaf 3 survived for one week. Leaf 3 needs only sunlight and water to live one week unattached from the plant.

*Experiment 10: Take Away the Light* 43

> • *All of the leaves without sunlight turned yellow-brown, but did not die.*
>
> (Answers will vary.)

Have the students discuss what they learned. Is it true that water and sunlight alone are sufficient for the survival of a detached leaf? How about for a leaf that remains attached, but lacks sunlight—is sunlight required for this leaf to survive?

## IV. Why?

Read this section of the *Laboratory Notebook* with your students.
Discuss any questions that might come up.

## V. Just For Fun

In this experiment students will grow bean plants to observe what happens when they are placed in locations with differing amounts of sunlight.

❶ Have the students decide how many plants they would like to grow for this experiment. Growing at least 4 plants will work best. They will be placing them in locations that get different amounts of sunlight. Have the students put potting soil in the pots, plant 3-4 bean seeds in each, and water them thoroughly.

❷ Have the students label each plant pot with a number or some other identifier they can reference when recording their observations.

❸ The pots should be placed in a warm, sunny location and the soil kept moist until the seeds sprout and begin to grow into seedlings. Depending on the variety of bean chosen, it should take about 7-14 days for the sprouts to show above the soil.

❹ Help the students find locations with varying amounts of sunlight. One plant should be placed in full sun and one where there is no sun, for example, in a closet. The other plants should be put in locations that have varying amounts of sunlight. The plants should be placed either all indoors or all outdoors and need to be watered regularly.

❺ Over the course of several weeks the students will observe the plants and record observations of their growth and health. Help the students think about what data they will need to record and then have them create a chart. Space is provided for recording their observations or they can use their field notebook or separate paper. Each plant should be listed by its identifier along with the amount of sunlight the plant will get during the day. Space should be provided for periodic notes made about the condition of the plant. Students might want to have a page for each plant, and they may want to draw as well as write.

At the end of the experiment, have the students draw conclusions based on their observations, then write them in the space provided. What did they learn about photosynthesis and sunlight?

# Experiment 11

# Seeing Inside Plants

### Materials Needed

- microscope with 4X, 10X, and 40X objective lenses; a 100X objective lens is recommended
- glass microscope slides (plain)
- glass microscope coverslips
- immersion oil (if using 100X objective lens)
- water
- eyedropper
- sharp knife
- toothpick
- colored pencils
- Samples:
  raw celery stalk with leaves
  raw carrot
  a large leaf
  other plant parts from plants of students' choice
- 3 or more small jars
- several fresh white carnation flowers
- food coloring

Microscope supplies can be purchased from Home Science Tools
https://www.homesciencetools.com/

## Objectives

In this experiment students will use a microscope to observe plant cells and tissues.

The objectives of this lesson are for students to:

- Observe how plants are made up of different cells and tissues that can be seen with a microscope.
- Understand that plants have different tissues and cells that perform different functions.

## Experiment

### I. Think About It

Read this section of the *Laboratory Notebook* with your students.

Ask questions such as the following to guide open inquiry.

> - *Which organs of a plant can you think of and what does each organ do?*
> - *Where and how do you think plants get nutrients?*
> - *What do you think would happen if a plant did not have xylem? Phloem?*
> - *What do you think you could observe if you looked at parts of plants with a microscope?*
> - *Why do you think a plant needs different kinds of cells?*
> - *Why do you think a plant needs different kinds of tissues?*

### II. Experiment 11: Seeing Inside Plants

Have the students read the entire experiment and then write an objective and a hypothesis.

#### Preparing Wet Mount Slides

Students are provided with the following instructions for preparing wet mount slides for this experiment.

The samples in this experiment should be prepared with wet slides, which are also called wet mounts. The water in a wet mount slide helps support the sample and can also help flatten it and make it more translucent. In addition, by filling the space between the coverslip and the slide, the water allows light to pass more easily through the sample.

A wet mount is made by simply putting a drop of water in the middle of a plain slide and gently placing the sample on top of the water. A coverslip is then placed over the sample by placing one edge of the coverslip at the edge of the water on the slide and carefully lowering the coverslip over the sample. A toothpick can be used for lowering the coverslip. Gently lowering the coverslip in this way will prevent air bubbles from forming under it. Having some air bubbles is okay, but too many will make it hard to see the sample.

If there is too much water on the slide and the coverslip won't stay in one place, you can hold a piece of paper towel at the edge of the slide and coverslip to draw out some of the water. If there is too little water under the coverslip, you can put a drop of water right next to the coverslip and some of the water will run underneath it. It takes a little practice to know how much water to use.

## EXPERIMENT

Note: If needed, have the students refer to Experiment 3 to review instructions for using the microscope.

In this experiment students will be looking through a microscope at sections of plant parts. The sections will need to be cut as thin as possible to allow enough light to pass through that the cellular structures can be observed. This requires use of a sharp knife. Students should be supervised while cutting the sections, or you may want to cut the sections yourself and then have the students prepare the slides.

❶ Supervise the students while they cut a very thin cross section of a celery stalk, or cut it for them. The slice needs to be as thin as possible so light can pass through.

❷ Have the students make a wet mount of the celery sample.

❸ Have the students place the sample in the microscope and begin by using the 4X magnification and then using successively higher magnifications.

The following instructions are included for using the 100X oil immersion lens:

*Remember that when you are using a 100X oil immersion lens, you need to turn the turret until the oil immersion lens is halfway into position. Place a single drop of immersion oil on the glass coverslip and gently move the oil immersion lens in place.*

*[NOTE: **It is extremely important that the lens does NOT scrape the coverslip. This can scratch the lens and ruin it.** The lens will be very close to the glass coverslip if you have adjusted the focus correctly with the other lenses. If it seems like the lens will scrape the coverslip, gently back up the lens with a few turns of the fine adjust knob.]*

Have the students carefully observe the celery sample and try to identify different kinds of cells and tissues. Have them refer to the *Student Textbook* for examples and descriptions.

❹ Space is provided in the *Results* section for students to make a drawing of their observations. Gathering the details for the drawing will help them make more careful observations. Have them label the types of cells and tissues in their drawing. Again, they can refer to the *Student Textbook* to help with identification.

*Experiment 11: Seeing Inside Plants* 47

Celery is a dicot. The vascular tissues are localized near the epidermal tissues and the pith makes up the central portion of the stem.

❺ Have the students repeat Steps ❶-❹ with a carrot. Since a thin slice of carrot can be more difficult to cut than one of celery, you may want to cut the section yourself.

❻ Space is provided in the *Results* section for students to compare and contrast the celery and carrot samples with drawings and/or writing. Help them think about the different functions of stems and roots. Can they see how these differences are reflected in the samples?

❼ Students will next look at the underside of a celery leaf. Have them experiment with making a dry slide of one leaf and a wet mount of another to see which works better. Ask them to discuss their observations about the slides.

Have the students repeat Steps ❸ and ❹ for the celery leaf. They will be looking for epidermal cells. Students may or may not notice stomata (singular, stoma), which are tiny openings on the underside of leaves. Plants use stomata to take in carbon dioxide and release oxygen. Are there other leaf structures they can notice, such as small veins? What can they notice by using the microscope that they cannot see with just their eyes? Ask them if they think they might be able to see more details if they had a thinner sample.

❽ Help the students prepare a sample from a large leaf. You or the students can use a sharp knife to cut a strip about 1" wide across the middle of the leaf. Roll the strip up and cut a few very thin slices from the roll. Have the students decide which slice they think will work best and then make a wet mount sample so they are looking at the cut edge of the leaf with the midvein of the leaf at the center of the slide. Have them record their observations in the *Results* section.

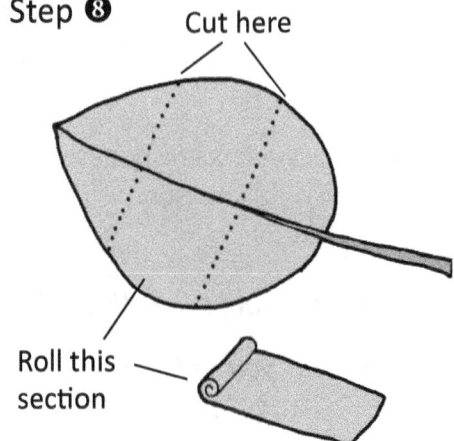

Step ❽

❾ Have the students gather samples from other plants and follow the appropriate instructions above to prepare slides and make observations. In addition to stems, roots, and leaves, they might like to look at parts of flowers, fruit, and seed pods. Plants can be edible or inedible.

Space for recording observations is provided in the *Results* section, or students can use their field notebook or separate sheets of paper.

## III. Conclusions

Have the students review the results they recorded for the experiment. Have them draw conclusions based on the data they collected.

## IV. Why?

Read this section of the *Laboratory Notebook* with your students.
Discuss any questions that might come up.

## V. Just For Fun — Colorful Flowers

In this experiment students will observe the transport of water through the xylem of carnation stems. Two variations of transport will be investigated. First, simple transport up the stem of the carnation will be examined. This part of the experiment serves to establish a control so that predictions about the second variation can be made. The transport observed in the first variation should result in petals with single colors. This experiment demonstrates that colored water is indeed transported up the stem to the petals of the flower.

In the second variation, the stem is split lengthwise starting about halfway up the stem and cutting away from the flower. Each of the divided stem ends is placed in a separate jar with a different color of water. The results will further demonstrate how water is transported through the flower.

Have the students read the entire experiment and then predict what they think will happen in each case. Ask questions such as the following to guide open inquiry:

- Do you think the colored water will travel to the petals of the flower?
- Do you think two different colors will travel up to the flower in the second part of the experiment?
- Do you think the petals on the flower in the second part will be a single color, or will the petals have two colors?
- Do you think the colored water will travel straight up in the second experiment, or will the colors mix in the stem? Will the colors mix in the flower to make a new color? Will they be intertwined? How do you think you will be able to tell what is happening?

Have the students read the entire experiment and then write an objective and a hypothesis.

## EXPERIMENT

❶ Have the students put water and several drops of food coloring in each of two small jars, using a different color in each jar and setting one of the jars aside. If desired, students can make additional jars of colored water to test several carnations.

❷ Help the students trim the end of a carnation stem at an angle and place it in one of the jars of colored water.

❸ Have them check the petals of the carnation every couple of hours, and record any color changes observed. More veins may show if left overnight.

❹ Once the flower is colored, have the students take the carnation out of the jar and help them cut a thin slice of the stem, make a wet mount slide, and look at it under the microscope. Have them try to identify the xylem, phloem, and other cells and tissues. Can they see epidermal cells? Ground tissue? There is space provided on the second page of the *Results* section for them to draw and label what they see. You can also have them cut the carnation flower lengthwise and look at parts of the flower under the microscope.

❺ Help the students slice a stem lengthwise with a knife, starting about halfway up the stem and cutting away from the flower, then stick one end of the divided stem into a jar of colored water and the other end in a jar with water of a different color. Have them let the carnation soak up the colored water until the petals begin to change color. In the *Results* section have them draw what they observe.

## Results

Space is provided for students to draw their results. In addition, petals can be removed from the flower and taped into the workbook. The petals will not be uniformly colored, but will have a dark strip of color at the very end. If enough time is allowed, the veins in the petals will also become colored.

The split carnation will result in a flower with two colors — one color on each side of the flower. The colored water travels up the split stem and colors only the petals on the same side as the piece of stem. The colors do not mix, and the split stem does not prevent the colored water from being transported.

## III. Conclusions

Have the students review the results they recorded for the experiment. Have them draw conclusions based on the data they collected.

# Experiment 12

# Growing Vegetables from Scraps

### Materials Needed

- several fresh vegetable scraps such as: carrot top, lettuce leaves or the root end of a head of lettuce, red beet top, turnip top, garlic bulb, onion bulb, scallions, either or both ends of a zucchini squash or cucumber, basil leaves with stem, potato (piece or peeling with eyes), or other vegetables of students' choice
- knife
- toothpicks
- several small glass jars or small drinking glasses
- colored pencils or pens
- several plant pots
- potting soil
- water

### Optional

- existing or new field notebook
- garden trowel or spoon

## Objectives

In this experiment students will explore vegetative reproduction by growing plants from cuttings.

The objectives of this lesson are for students to:

- Learn more about vegetative reproduction by direct observation.
- Observe how plants grow differently from different structures.

## Experiment

### I. Think About It

Read this section of the *Laboratory Notebook* with your students.
Ask questions such as the following to guide open inquiry.

- *Why do you think there are so many more species of vascular plants than nonvascular plants?*
- *Why do you think flowering plants are so successful that they make up about 80% of all plant species?*
- *Do you think if you cut a vegetable into several pieces, the pieces would grow? Why or why not?*
- *If you cut a vegetable into several pieces and tried to grow a plant from one of the pieces, do you think it would matter which piece you chose? Why or why not?*
- *If you tried to grow a vegetable plant from a cutting, what conditions do you think you would need to provide for the cutting? Why?*
- *Do you think different types of plants would need different growing conditions? Why or why not?*

### II. Experiment 12: Growing Vegetables from Scraps

Have the students read the entire experiment before writing an objective and a hypothesis.

**Objective:** Have the students write an objective. (What will they be learning?)

**Hypothesis:** Have the students write a hypothesis.

## EXPERIMENT

**❶** In this experiment students will explore vegetative reproduction by trying to grow vegetable plants from cuttings, using vegetable parts that are usually discarded during meal prep. Help the students decide which vegetables and what part of the vegetable they would like to test. They are given these suggestions:

Carrot top, lettuce leaves or root end of a head of lettuce, red beet top or root, turnip top or root, garlic bulb, onion bulb, scallions, either or both ends of a zucchini squash or cucumber, basil leaves with stem, potato (piece or peeling), other vegetables of your choice. (If the students want to try growing a potato, an organic one will probably grow better because conventionally grown potatoes are treated with a growth inhibitor. Guide students to the observation that for the potato piece to grow, one or more eyes are needed.)

If there are leaves on a vegetable top, for example a carrot or red beet, these should be cut off.

Have the students look at each vegetable before and after it is cut up, and discuss the different plant structures and what function they perform. Have them decide which part of the plant is most likely to grow.

Help them cut off each vegetable piece they would like to grow. Have them choose different vegetables that have different parts that may grow (tops, bulbs, etc.) and they can also test vegetable pieces that they are not sure will grow. The cutting will need to be about 1.5-3 cm (.5-1 inch) thick. Have them cut the vegetables just before doing the experiment.

**❷** Have them gather enough small glass jars or drinking glasses to hold the vegetable cuttings.

**❸** The cuttings will need to be suspended from the top of a jar so they are in water but not submerged. Have the students decide which side of the cutting should go down into the water. Guide them to thinking about roots growing down and stems growing up. Then have them put enough toothpicks into the sides of the cutting that when the toothpicks are resting on the jar rim, they will suspend the cutting.

**❹** The cuttings will need to be placed in a warm, sunny location, then water can be added to the jars. Each cutting should have water covering the bottom and going partway up the sides.

**❺** Have the students check the jars daily to make sure the water is still covering the bottom of the cuttings. The water can be changed occasionally if needed.

**❻** Students are asked to observe their cutting samples over the course of several weeks and record observations in the *Results* section or in their field notebook.

### Results

Space is provided for students to write and draw their observations. They are asked to make a section for each plant's data and note the type of plant, the part of the plant, and the date the cutting is put in water. They are also asked to make a prediction about whether or not the cutting will grow and why they think this. They will be observing the cuttings daily

*Experiment 12: Growing Vegetables from Scraps*    53

as they check the water levels, and they are asked to record any changes they see in the cuttings, along with the date of the observation. They are asked to record their observations both in writing and with drawings. They are instructed to fasten more sheets of paper into this section of the workbook if they run out of space for recordation.

## III. Conclusions

Have the students review the results they recorded for the experiment. Have them draw conclusions based on the data they collected. Have them think about how easy or difficult it was to grow plants from cuttings, what problems they ran into, whether the cuttings grew or not, and things they might do differently to get different results.

## IV. Why?

Read this section of the *Laboratory Notebook* with your students.
Discuss any questions that might come up.

## V. Just For Fun

### Eating Scraps

Most of the vegetable scraps listed in the experiment can grow roots, stems, and leaves. Some can grow into vegetables that can be eaten. For example, lettuce can grow leaves that are big enough to eat. Carrot leaves are edible and recipes can be found on the internet.

**Note:** Potato stems, leaves, and roots contain a toxin called solanine and should not be eaten. Also potatoes that have been exposed to light and are green below the skin have a build up of this toxin and should not be eaten. Potatoes that are not green are safe to eat.

When students determine a cutting has grown enough roots to be transplanted, they are to plant it in a pot with potting soil and then put the plant in a warm, sunny location They will need to check the pots often to make sure the soil stays moist.

Have the students record the growth and health of their plants. They are asked to note whether the plants get big enough to harvest vegetables to eat, and if so, how long it took. Also have them note which cuttings grow best and which ones die or don't grow well. Ask them why they think they got these results.

# Experiment 13

# Human Cheek Cells

**Materials Needed**
- toothpicks or cotton swabs
- glass microscope slides*
- plastic pipette or eyedropper*
- methylene blue solution (0.5% to 1%)* (iodine can be used instead—follow the same safety precautions)
- plastic cover slip*
- paper towels or tissues
- thin rubber, vinyl, or latex gloves that are a tight fit
- goggles or other eye protection*
- microscope

**Optional**
- immersion oil*

\* Available from Home Science Tools: https://www.homesciencetools.com/

Type the name of the item needed in the website search bar.

## Objectives

In this experiment students use a microscope to look at cells from their cheek.

The objectives of this lesson are for students to:

- Explore the structure of one type of human cell.
- Use a microscope to observe cells.

## Experiment

### I. Think About It

Read this section of the *Laboratory Notebook* with your students.

Ask questions such as the following to guide open inquiry.

- *What methods do you think scientists use to study cells?*
- *What do you think you could find out about cells by using a microscope?*
- *Why do you think a scientist might use a dye when studying cells?*
- *Do you think you could study some of your own cells? How would you do it? What cells would you choose? Why?*

### II. Experiment 6: Human Cheek Cells

Have the students read the entire experiment before writing an objective and a hypothesis.

**Objective:** Have the students write an objective based on what they think they will be learning.

**Hypothesis:** Have the students write a hypothesis—what they think will happen.

### EXPERIMENT

In this experiment students will use a microscope to observe cells from the inside of their mouth.

❶ Make sure the students use goggles or other eye protection and protective rubber, vinyl, or latex gloves that are tight-fitting.

❷ Have them use a clean cotton swab or toothpick and gently scrape the inside of their cheek to collect cells.

❸ Students will rub the cotton swab or end of the toothpick on the center of a microscope slide for 2–3 seconds.

❹ Have them carefully add a drop of methylene blue solution (or iodine) on the sample that is on the slide and place a coverslip on top. [CAUTION: Both methylene blue and iodine are toxic if ingested. Direct students to wear gloves and goggles when using the dye and do NOT let them swallow the chemical or put their fingers in or near their mouth while handling it. Have them wash their hands and surfaces after use.]

❺ Students should avoid touching the coverslip while carefully looking for tiny air bubbles under it. Air bubbles can interfere with observation of the cheek cells. Have the students take a paper towel and gently touch it to the corner of the coverslip to absorb any excess solution. If they are still seeing air bubbles, have them add a little more liquid at the side of the cover slip and very gently press down on the cover slip. If there are still bubbles, have them start over.

❻ Students will set up their microscope by moving the 4X or 10X objective lens into position and placing the slide in the microscope. To examine a cell in more detail, have students move to a higher objective (40x—or 100x if they have an oil immersion lens).

❼ In the Results section, have the students record their observations by drawing what they see.

# Results

Space for students to record observations is found here. They may want to use additional paper.

## III. Conclusions

Have the students review the results they recorded for the experiment. Have them draw conclusions based on their observations. How easy or difficult is it to examine human cells. Why? What did they discover about these cells?

## IV. Why?

Read this section of the *Laboratory Notebook* with your students.
Discuss any questions that might come up.

## V. Just For Fun

## Make Your Own Microscope Dye

Students will explore making dyes to use with their microscope samples.

They can test a variety of household materials, such as red or blue food coloring, iodine, India ink, red cabbage juice, pomegranate juice, or juice from other deeply colored plants. They can learn more by doing research on the internet or at the local library.

Students can also try gathering cells from their parents, dog, horse, or other animals patient enough to let their inner cheek be gently scraped. Have them compare their cheek cell to these other cells.

# Experiment 14

## Non-chordates

### 14A: Gross Anatomy Comparison Using Dissection
### 14B: Picky Eaters?

**Students have the choice of one of two experiments, or they can choose to do both**

There are two options available for this experiment. In Experiment A students will dissect purchased samples to learn about internal structures of animals. In Experiment B students will observe the behavior of live animals.

Either option will help students learn more about non-chordates. Which option is chosen is just a matter of preference and depends on what the student wants to focus on. Some students enjoy dissection, have an interest in anatomy, and want to explore how different animal bodies work on the inside. Other students are more interested in animal behaviors and environments. Some students will be interested in performing both experiments.

# Experiment 14A
# Gross Anatomy Comparison Using Dissection

**Materials Needed**

- preserved non-chordate organisms, non-injected specimens are OK (clam, crayfish, sea star, and earthworm)
- a dissection guide for each organism
- safety goggles
- lab apron
- gloves
- dissection tray
- dissection pins
- dissecting probe
- forceps
- scissors
- scalpel
- hand lens or magnifying glass
- paper towels
- water

Preserved organisms and dissection guides are available from Home Science Tools. On the website, search for "dissection specimen" and "dissection guide." Choose the four organisms listed above. (At the time of this writing, Home Science Tools offers an "Animal Specimen Set of 9 with Pig" that has most of the specimens needed for Experiments 14-16)

Dissection tools are also available from Home Science Tools. Search for individual tools or a dissection kit. Other supplies also available.

https://www.homesciencetools.com/

*Experiment 14: Non-chordates*  59

## Objective

In this experiment students will observe the external and internal structures of four different non-chordates. The objectives of this lesson are for students to:

- Use dissection as a method for studying non-chordate body structures.
- Observe, compare, and contrast body structures of four different non-chordate animals.

## Experiment

### I. Think About It

Read this section of the *Laboratory Notebook* with your students.
Ask questions such as the following to guide open inquiry.

- *What do you think you can learn about non-chordates by studying the outside of their bodies?*
- *What external body structures would you expect to find?*
- *What do you think you can learn about non-chordates through dissection?*
- *What internal body structures would you expect to find?*
- *What do you think you would learn by dissecting different types of non-chordates?*

### II. Experiment 14A: Gross Anatomy Comparison Using Dissection

Have the students read the entire experiment before writing an objective and hypothesis.

**Objective:** An objective is provided.

**Hypothesis:** This is an observational experiment so there is no hypothesis.

### EXPERIMENT

In this experiment students will dissect four non-chordate specimens.

❶ Have the students review the dissection guide for the specimen.

❷ Have students put on safety goggles, gloves, and apron.

❸ Students will take one of the preserved non-chordate organisms, use water to rinse off any excess preservative, and then place it on the dissection tray.

❹ In one of the boxes provided in the *Results* section, students will write the name of the type of organism being observed (clam, crayfish, sea star, or earthworm).

❺ Have students observe the top, bottom, and sides of the organism and note whether the organism has radial or bilateral symmetry. (In bilateral symmetry, the organism can be divided in half lengthwise, and both halves will have the same features, but it can't be further divided and still have symmetrical parts.) In the box in the *Results* section, have the students write down the type of symmetry and draw the top, bottom, and side (or sides) of the animal.

❻ Have the students identify the ventral side of the organism and label it "ventral" in their drawing. (The ventral surface is the lower or underside of the body.)

❼ Have the students identify the dorsal side of the organism and label it "dorsal" in their drawing. (The dorsal surface relates to the upper side or back of the body.)

❽ Using the hand lens or magnifying glass, students will examine the outer surface of the organism and note any external structures that can be seen. These can include but are not limited to bristle-like structures called setae (earthworm); sharp spines or tube feet (sea star); antennae, eyes, claws, jointed feet (crayfish); etc. Have the students label these features on their drawing.

❾ Have the students place the organism either dorsal side up or dorsal side down according to the instructions for dissection and then begin dissecting, using the scalpel and scissors. In the box in the *Results* section for this animal, have students draw what they see, noting the various internal structures, including but not limited to the heart, gut, and reproductive organs.

❿ Students will repeat the experiment with the second, third, and fourth organisms.

# Results

Space for students to record observations is found here. They may want to use additional paper.

## III. Conclusions

Have the students review the results they recorded for the experiment. Have them draw conclusions based on their observations. How easy or difficult is it to learn about the internal structures of animals through dissection? Why? What did they discover about non-chordates?

## IV. Why?

Read this section of the *Laboratory Notebook* with your students and discuss any questions.

## V. Just For Fun

# Observing Tissues and Cells

Have the students use their microscope to take a closer look at tissue samples from the organisms they dissected. Have them use a low magnification lens (10X) and draw what they see.

# Experiment 14B
# Picky Eaters?

**Materials Needed**
- small ants
- food items:
  sugar cube
  small piece of animal protein (chunk of turkey, ham, roast beef, etc.)
  cheese
  apple
  bread
  oil or butter
- homemade choice chamber: a shallow pan, a shallow cardboard box, a short jar, or a plastic Petri dish;
  cardboard or paper cut into strips;
  tape or glue
- purchased choice chamber: an inexpensive one is available from Home Science Tools; search on "choice chamber."
  https://www.homesciencetools.com/

## Objective

In this experiment students will observe the food preferences of ants.

The objectives of this lesson are for students to:

- Observe animal behavior.
- Use a simple tool.

## Experiment

### I. Think About It

Read this section of the *Laboratory Notebook* with your students.
Ask questions such as the following to guide open inquiry.

> - *What types of animal behaviors do you think you can observe? How would you observe them? Where?*
>
> - *Do you think you could do a test for animals' food preferences? If so, how would you do it? What animals do you think you could test for food preferences?*
>
> - *Do you think you would need to design different food preference tests for different animals? Why or why not?*
>
> - *Do you think all species of ants will have the same food preferences? Why or why not?*

### II. Experiment 14B: Picky Eaters?

Have the students read the entire experiment.

**Objective:** Have the students write an objective based on what they think they will be learning.

**Hypothesis:** Have the students write a hypothesis.

### How To Make a Choice Chamber

Have the students take a shallow pan, a shallow cardboard box, a short jar, or a plastic Petri dish. Have them use thin strips of cardboard or paper to divide the container into 2, 3, or 4 sections. To do a test they will place the animals in the center and observe their movements.

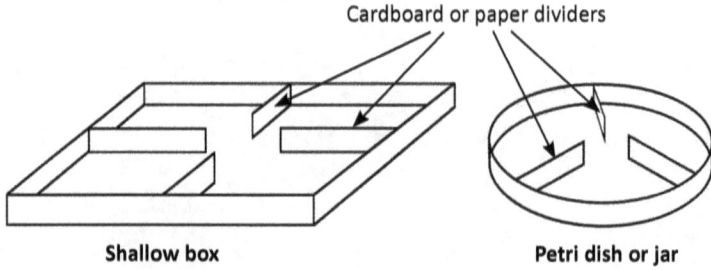

Shallow box   Petri dish or jar

## EXPERIMENT

Students will perform a simple experiment to observe animal behavior.

❶ Have the students collect some small ants in a jar. Have them look for the ants near your house, in a park, or in an empty field.

❷ Have the students set up the choice chamber. A choice chamber may be purchased or the students can make their own.

❸ Students will place a sugar cube on one side of the choice chamber and meat on the other.

❹ Help them carefully pour or place the ants in the chamber and close the lid.

❺ Have the students observe the ants. After the ants have been in the choice chamber for about an hour, have the students record which food item the ants have chosen. They will record their observations in the chart in the Results section. If both food items are equally populated, have them record both in the chart.

❻ Have the students repeat the experiment one or more times using different food combinations.

❼ When the students have finished the experiment, have them return the ants to the place where they were found.

## Results

A chart for students to record their observations is found here.

### III. Conclusions

Have the students review the results they recorded for the experiment. Have them draw conclusions based on their observations. How easy or difficult is it to study animal behavior? Why? What did they discover?

### IV. Why?

Read this section of the *Laboratory Notebook* with your students.
Discuss any questions that might come up.

### V. Just For Fun

## More Choices!

Have the students design their own choice test. They could try the food choice test with a different animal, like a snail or pill bug. Or they might choice test for light, water, odor, or some other factor. Encourage them to use their imagination in thinking of additional tests.

# Experiment 15

## Chordates

### 15A: Gross Anatomy Comparison Using Dissection
### 15B: Bird Migration

### Students have the choice of one of two experiments

There are two options available for this experiment. Students who choose Experiment A will dissect purchased specimens to learn about internal structures of animals. Students who choose Experiment B will observe the behavior of live animals.

Either option will help students learn more about chordates. Which option is chosen is just a matter of preference and depends on what the student wants to focus on. Some students enjoy dissection, have an interest in anatomy, and want to explore how different animal bodies work on the inside. Other students are more interested in animal behaviors and environments.

# Experiment 15A
# Gross Anatomy Comparison Using Dissection

**Materials Needed**
- preserved chordate organisms (frog, shark, and perch). Specimens don't need to be injected.
- dissection guide for each organism
- safety goggles
- lab apron
- gloves
- dissection tray
- dissection pins
- dissecting probe
- forceps
- scissors
- scalpel
- hand lens or magnifying glass
- paper towels
- water

Preserved organisms are available from Home Science Tools. On the website, search for "dissection specimen" and choose the three organisms listed above. (At the time of this writing, Home Science Tools offers an "Animal Specimen Set of 9 with Pig" that has most of the specimens needed for Experiments 14-16.)

Dissection tools and guides are also available from Home Science Tools. Search for individual tools or a dissection kit and for dissection guides.

https://www.homesciencetools.com/

## Objective

In this experiment students will observe the external and internal structures of three different chordates.

The objectives of this lesson are for students to:

- Use dissection as a method of studying chordate body structures.
- Observe, compare, and contrast body structures of three different chordate animals.

## Experiment

### I. Think About It

Read this section of the *Laboratory Notebook* with your students.

Ask questions such as the following to guide open inquiry.

- *What do you think you can learn about chordates by studying the outside of their bodies?*
- *What external body structures would you expect to find?*
- *What do you think you can learn about chordates through dissection?*
- *What internal body structures would you expect to find?*
- *How do you think the external structures of chordate bodies will compare to those of the non-chordates you examined in the last experiment?*
- *How do you think the internal structures of chordate bodies will compare to those of the non-chordates you dissected in the last experiment?*

### II. Experiment 15A: Gross Anatomy Comparison Using Dissection

Have the students read the entire experiment before writing an objective.

**Objective:** An objective is provided.

**Hypothesis:** This is an observational experiment so there is no hypothesis.

# EXPERIMENT

In this experiment students will dissect three different chordate specimens.

❶ Have students review the dissection guide for the specimen to be dissected.

❷ Have students put on safety goggles, gloves, and apron.

❸ Students will take one of the preserved chordate organisms, use water to rinse off any excess preservative, and then place it on the dissection tray.

❹ In one of the boxes provided in the *Results* section, students will write the name of the type of organism being observed (frog, shark, or perch).

❺ Have students observe the top, bottom, and sides of the organism and note whether the organism has radial or bilateral symmetry. (In bilateral symmetry, the organism can be divided in half lengthwise, and both halves will have the same features, but it can't be further divided and still have symmetrical parts.) In the box in the *Results* section, have the students write down the type of symmetry and draw the top, bottom, and side (or sides) of the animal.

❻ Have the students identify the ventral side of the organism and label it "ventral" in their drawing. (The ventral surface is the lower or underside of the body.)

❼ Have the students identify the dorsal side of the organism and label it "dorsal" in their drawing. (The dorsal surface relates to the upper side or back of the body.)

❽ Using the hand lens or magnifying glass, students will examine the outer surface of the organism and note any external structures that can be seen. These can include but are not limited to scales, bumps, and coloration. Have the students label these features on their drawing.

❾ Have the students place the organism either dorsal side up or dorsal side down according to the instructions for dissection and then begin dissecting, using the scalpel and scissors. In the box in the *Results* section for this animal, have students draw what they see, noting the various internal structures, including but not limited to the heart, gut, and lungs.

❿ Students will repeat the experiment with the second and third organisms.

# Results

Space for students to record observations is found here. They may want to use additional paper.

## III. Conclusions

Have the students review the results they recorded for the experiment. Have them draw conclusions based on their observations. How easy or difficult is it to learn about the internal structures of animals through dissection? Why? What did they discover about chordates?

## IV. Why?

Read this section of the *Laboratory Notebook* with your students. Discuss any questions that might come up.

## V. Just For Fun

# Observing Tissues and Cells

Students will use their microscope to take a closer look at tissue samples from the organisms they dissected. Have them use a low magnification lens (10X) to observe details that can't be seen with eyes alone. Have them record their observations on separate paper.

# Experiment 15B
# Bird Migration

**Materials Needed**
- ebird.org app (free)
- Merlin Bird ID app (free) or other bird ID app and/or a print book field guide to the birds, such as *The Young Birder's Guide to North America*
- smartphone or iPad with internet access and camera; or desktop or laptop computer and digital camera, if available
- an email address
- field notebook (an existing one or start a new one for birds)
- pen, pencil, colored pencils

**Optional**

binoculars

## Objective

In this experiment students will participate in a Cornell University citizen science project to track bird populations and migrations.

The objectives of this lesson are for students to:

- Learn about citizen science.
- Observe and identify birds in the area.

## Experiment

### I. Think About It

Read this section of the *Laboratory Notebook* with your students.

Ask questions such as the following to guide open inquiry.

> - *How many different kinds of birds do you think you see in a day?*
> - *How many different kinds of birds can you identify and name? Identify by appearance? By song?*
> - *Do you see birds all year where you live? If so, which ones? Is there a time of the year when you see the most birds?*
> - *When birds migrate (move from one region to another according to the season), how do you think they know where to go?*
> - *Why do you think different birds live in different areas?*

### II. Experiment 15B: Bird Migration

Have the students read the entire experiment.

**Objective:**   Have the students write an objective based on what they think they will be learning.

**Hypothesis:**  Have the students write a hypothesis. For example:

> There will be a seasonal pattern for some birds.
> Some birds migrate only a few hundred miles.
> Some birds migrate thousands of miles.
> Some birds migrate between countries, like between the U.S. and Canada.

## EXPERIMENT

Students will take part in a citizen science project by identifying birds in their area and submitting their observations to Cornell Lab's eBird database. The data is used by scientists to track bird populations and migrations.

❶ Students will go to the ebird.org website.

❷ Have them click on *Learn More* and watch the 3 minute video.

❸ Help the students create a Cornell Lab account.

❹ Confirmation of their account will go to their email inbox.

❺ Have them read the privacy policy and terms of use and then help them decide which settings they prefer.

❻ If the students have a mobile device, have them download the Merlin Bird app, a free app for smart phones and iPads that will help with bird identification. On the website, they will click on the *Explore* tab, scroll down to Merlin Bird ID, and download the app. If they don't have a mobile device, they can use a field guide book such as *The Young Birder's Guide to North America*. Or they can use both.

❼ Have them go back to the *Explore* tab and provide their county and state (or other information if outside the U.S.) in the *Explore Regions* box. This will bring up a list of birds sighted in their area. Clicking on the name of the bird brings up a picture of it. They can choose a bird from this list, or they can observe the birds near their home or school and identify them. If they have a mobile device, they can find out what type of bird has been spotted by taking a photo of it and using the Merlin Bird ID app.

Have them spend some time exploring the eBird website.

❽ Take the students outside to look for birds

❾ In their field notebook, have them record the birds they see. Have them identify the type of bird and note the date, time, location, weather, and season. Have them take a photo of each bird and upload it to their eBird account using the *Submit* button.

Have them draw the bird or place a photo of it in their field notebook.

❿ Have them continue collecting data, submitting it to eBird, and recording it in the field notebook. How many different birds can they identify?

## Results

Students will record their observations in their field notebook.

### III. Conclusions

Have the students go to ebird.org, click on the *Science* tab, and locate the trending movement and location for one or more of the birds they've identified. Does their observation correlate to the abundance map? Have them explain why they think it does or does not. What did they learn about birds in their area?

### IV. Why?

Read this section of the *Laboratory Notebook* with your students.
Discuss any questions that might come up.

### V. Just For Fun

## Follow That Bird!

Students are provided with the following instructions. They can pick one experiment or do both.

- Choose several species of birds to observe over time, and put daily or weekly observations in your field notebook. How many different things can you observe about each bird? For example, where it lives, what it eats, whether it spends most of its time by itself or with others, what kind of song it has, where it nests. Did it have babies while you were observing it? Does it stay in your area all year? Do the males and females look the same? Ask as many question as you can think of and see if you can find the answers. You can also make a chart to compare data about different birds.

- To learn lots more about birding and eBird, go to ebird.com and click on the *Explore* tab. Scroll down to *Learn Birding Skills*. Follow the directions to sign up for the free *eBird Essentials* course and get started. It has stuff to read and videos to watch. It's self-directed, so you can go at your own speed.

# Experiment 16

# Mammals

**16A: Gross Anatomy Using Dissection**
**16B: Pick Your Citizen Science Project!**

## Students have the choice of one of two experiments

There are two options available for this experiment. Students who choose Experiment A will dissect a purchased preserved fetal pig specimen to learn about internal structures of a mammal. Students who choose Experiment B will participate in the citizen science. project of their choice.

# Experiment 16A
# Gross Anatomy Using Dissection

**Materials Needed**

- preserved fetal pig (doesn't need to be injected)
- dissection guide
- safety goggles
- lab apron
- gloves
- dissection tray
- dissection pins
- dissecting probe
- forceps
- scissors
- scalpel
- hand lens or magnifying glass
- paper towels
- water

Preserved fetal pig specimens are available from Home Science Tools. On the website, search for "dissection specimen."

Dissection tools and guides are also available from Home Science Tools. Search for individual tools or a dissection kit and for dissection guides.

https://www.homesciencetools.com/

## Objective

In this experiment students will observe the external and internal structures of three different chordates.

The objectives of this lesson are for students to:

- Use dissection as a method of studying mammalian body structures.
- Observe, compare, and contrast body structures of animals dissected in Experiments 14-16.

## Experiment

### I. Think About It

Read this section of the *Laboratory Notebook* with your students.

Ask questions such as the following to guide open inquiry.

- *What do you think you can learn about mammals by studying the outside of their bodies?*
- *What external body structures would you expect to find?*
- *What do you think you can learn about mammals through dissection?*
- *What internal body structures would you expect to find?*
- *How do you think the external structures of all the non-chordate and chordate animals you dissected will compare to each other?*
- *Do you think all the chordate animals you dissected will have similar internal organs?*

### II. Experiment 16A: Gross Anatomy Comparison Using Dissection

Have the students read the entire experiment before writing an objective.

**Objective:** Have the students write an objective based on what they think they will be learning.

**Hypothesis:** This is an observational experiment so there is no hypothesis.

## EXPERIMENT

In this experiment students will dissect a preserved fetal pig.

❶ Have students review the dissection guide for the fetal pig dissection.

❷ Have students put on safety goggles, gloves, and apron.

❸ Students will take the preserved fetal pig, use water to rinse off any excess preservative, and then place it on the dissection tray.

❹ In the box provided in the *Results* section, students will write the name of the specimen being observed (fetal pig).

❺ Have students observe the top, bottom, and sides of the fetal pig and note whether it has radial or bilateral symmetry. (In bilateral symmetry, the organism can be divided in half lengthwise, and both halves will have the same features, but it can't be further divided and still have symmetrical parts.) In the box in the *Results* section, have the students write down the type of symmetry and draw the top, bottom, and side (or sides) of the fetal pig.

❻ Have the students identify the ventral side of the organism and label it "ventral" in their drawing. (The ventral surface is the lower or underside of the body.)

❼ Have the students identify the dorsal side of the organism and label it "dorsal" in their drawing. (The dorsal surface relates to the upper side or back of the body.)

❽ Using the hand lens or magnifying glass, students will examine the outer surface of the fetal pig and note any external structures that can be seen. These can include but are not limited to hooves, snout, tail, and ears. Also have them note features such as variations in the skin in different areas of the body, coloration, etc. Have them label the features on their drawing.

❾ Have the students place the organism either dorsal side up or dorsal side down according to the instructions for dissection and then begin dissecting, using the scalpel and scissors. In the box in the *Results*, have students draw what they see, noting the various internal structures, including but not limited to the heart, gut, and lungs.

## Results

❶ Space for students to record observations is found here. They may want to use additional pieces of paper.

❷ Have the students answer the questions in this section.

## III. Conclusions

Have the students review the results they recorded for the experiment. Have them draw conclusions based on their observations. How easy or difficult is it to learn about the internal structures of animals through dissection? Why? What did they discover about mammals?

## IV. Why?

Read this section of the *Laboratory Notebook* with your students. Discuss any questions that might come up.

## V. Just For Fun

# Observing Tissues and Cells

Students will use their microscope to take a closer look at tissue samples from the fetal pig they dissected. Have them use a low magnification lens (10X) to observe details that can't be seen with eyes alone. Have them record their observations on separate paper.

# Experiment 16B
# Pick Your Citizen Science Project!

**Materials Needed**

- smartphone, iPad, or computer with internet access and camera; or desktop or laptop computer and digital camera, if available
- an email address
- field notebook (an existing one or start a new one for citizen science projects)

**Or**

- Local library, zoo, or natural history museum
- field notebook (an existing one or start a new one for citizen science projects)

## Objective

In this experiment students will learn more about citizen science by choosing the project they would like to be involved in.

The objectives of this lesson are for students to:
- Become a part of citizen science.
- Think about what areas of biology they would like to study further.

## Experiment

### I. Think About It

Read this section of the *Laboratory Notebook* with your students.

Ask questions such as the following to guide open inquiry.

> - *How do you think citizen science projects help scientists?*
> - *What kinds of citizen science projects would be most interesting to you?*
> - *Do you think there are citizen science projects being conducted in the area where you live? How would you find out?*
> - *Do you think you could design your own citizen science project? If so, how would you do it?*

### Experiment 16B: Pick Your Citizen Science Project!

Have the students read the entire experiment.

## EXPERIMENT

Students will do research to find a citizen science project they'd like to participate in. There are many different citizen science projects that will help students become involved in scientific research.

If students have internet access, help them do a search for citizen science projects and choose one they would enjoy. Some interesting sites to try are:

- scistarter.org
- thinking animalsunited.org
- aza.org
- inaturalist.org
- zooniverse.org

If your students do not have easy access to a computer or smartphone, they can go to the local library, zoo, or natural history museum and ask what local citizen science projects are being conducted in your area.

Have students use their field notebook to record observations.

### V. Just For Fun

## Create Your Own Citizen Science Project!

Students are invited to create their own citizen science project. They are directed to go outside and observe living things and environments to find something they'd like to learn more about. Then they will design their own citizen science project to collect data about this area of interest.

If needed, help them decide on a project. Have them write down details in their field notebook, such as: specifically what they want to find out, what data needs to be collected, who will collect the data, how it will be analyzed, etc. They can add details to this page as the project develops.

Have the students ask friends and family to participate in their citizen science project. Students will need to write a description of the project, the type of data that they want collected, and possibly a form to be filled out. When the data has been received from the participants, help the students decide how to organize and analyze it. Have them record the results and their observations and conclusions about their citizen science project in their field notebook.

# More REAL SCIENCE-4-KIDS Books
## by Rebecca W. Keller, PhD

**Building Blocks Series** yearlong study program — each Student Textbook has accompanying Laboratory Notebook, Teacher's Manual, Lesson Plan, Study Notebook, Quizzes, and Graphics Package

Exploring Science Book K (Activity Book)
Exploring Science Book 1
Exploring Science Book 2
Exploring Science Book 3
Exploring Science Book 4
Exploring Science Book 5
Exploring Science Book 6
Exploring Science Book 7
Exploring Science Book 8

**Focus On Series** unit study program — each title has a Student Textbook with accompanying Laboratory Notebook, Teacher's Manual, Lesson Plan, Study Notebook, Quizzes, and Graphics Package

Focus On Elementary Chemistry
Focus On Elementary Biology
Focus On Elementary Physics
Focus On Elementary Geology
Focus On Elementary Astronomy

Focus On Middle School Chemistry
Focus On Middle School Biology
Focus On Middle School Physics
Focus On Middle School Geology
Focus On Middle School Astronomy

Focus On High School Chemistry

## Super Simple Science Experiments

21 Super Simple Chemistry Experiments
21 Super Simple Biology Experiments
21 Super Simple Physics Experiments
21 Super Simple Geology Experiments
21 Super Simple Astronomy Experiments
101 Super Simple Science Experiments

**Note:** A few titles may still be in production.

## Gravitas Publications Inc.
www.gravitaspublications.com
www.realscience4kids.com

CPSIA information can be obtained
at www.ICGtesting.com
Printed in the USA
JSHW071956270323
39291JS00004B/24